AF775956

Without Empat

Without Empathy

Irony and the Satirical Impulse in Eight Major Filmmakers

MK Raghavendra

intellect
Bristol, UK / Chicago, USA

First published in the UK in 2025 by
Intellect, The Mill, Parnall Road, Fishponds, Bristol, BS16 3JG, UK

First published in the USA in 2025 by
Intellect, The University of Chicago Press, 1427 E. 60th Street,
Chicago, IL 60637, USA

Copyright © 2025 Intellect Ltd

All rights reserved. No part of this publication may be reproduced,
stored in a retrieval system, or transmitted, in any form or by
any means, electronic, mechanical, photocopying, recording, or
otherwise, without written permission.

No part of this book may be used or reproduced in any manner for the
purpose of training artificial intelligence technologies or systems
without written permission from the publisher.

A catalogue record for this book is available from
the British Library.

Copy editor: MPS Limited
Cover designer: Tanya Montefusco
Cover image: *Children's Games*, Pieter Bruegel The Elder
Production managers: Sophia Munyengeterwa & Tian Greaves
Typesetter: MPS Limited

Hardback ISBN 978-1-83595-140-8
ePDF ISBN 978-1-83595-142-2
ePUB ISBN 978-1-83595-141-5

To find out about all our publications, please visit our website.
There you can subscribe to our e-newsletter, browse or download our current
catalogue and buy any titles that are in print.

www.intellectbooks.com

To

Deep Focus, *the film quarterly,*
and all those who, with me alongside, launched it in January 1988:
MU Jayadev, Georgekutty AL, Babu Subramanian, Gerard
Sequeira, John Bhaktul and others

I have never made but one prayer to God, a very short one: Oh Lord, make my enemies ridiculous. And God granted it.
—Voltaire

Censorship is the mother of metaphor.
—Jorge Luis Borges

Contents

Acknowledgments

My interest in the films and filmmakers discussed in this book was stoked through the Saturday screenings with the film study group organized by Naresh Narasimhan at Venkataramanan Associates. I would therefore like to acknowledge members of the group, past and present, including Naresh, Sridhar, Jaidev, Hameed, Babu, Aditya, Suhas, Vishwanath, Romit, Devapriya, Basav, Kupendra, and Rudransh. The anonymous readers who scrutinized this book and helped me bring it to the present state contributed enormously through their invaluable suggestions and they must be acknowledged most of all.

Introduction: Irony, Satire, and Empathy

Background

This is a book focused on irony and the satirical impulse as expression in cinema through its appearance in the work of eight selected filmmakers of importance to cinema. By "irony" what is meant is neither situational irony or dramatic irony but only verbal irony or its cinematic equivalent (the treatment implying that something is not to be taken literally). Situational irony as a literary technique used in Greek tragedy pertains to a state of affairs or an event that seems deliberately contrary to what one expects. But there are different kinds of irony, many not even appropriate to satire: in Socratic irony, the speaker assumes a pose of ignorance and an eagerness to be instructed, but with persistent questioning the questions prove unanswerable; courtroom dramas often use this. Dramatic irony is when the spectator knows more than the character knows and in most cases the character's acts are grossly inappropriate to the situation as in an anti-war film like Stanley Kubrick's *Paths of Glory* (1957). Tragic irony would be used in a tale where the outcome is already known as in the retelling of epics (*Troy*, 2004) where a hero doomed to die goes off to his death, full of hope. Irony is a useful strategy to use when the theme dictates a certain emotional response, say an emotional one. Bertolt Brecht used irony extensively in plays like *Mother Courage and Her Children* (1939).

The book is primarily dedicated to artists in cinema who consistently use irony as an element, who have made films often too ambiguous to be termed "satire" but who nonetheless exhibit the impulse consistently over a large number of films. Irony and satire are commonly used to say something contrary to the truth in order for the truth to be exposed to the general public for the purpose of awareness and change, at least through interrogating dominant viewpoints. Often, these artists say things with a straight face that they want us not to accept. What marks out the films of the chosen filmmakers here is that all of them say something they do not actually mean and expect people to understand not only what they actually do mean but also their attitude toward it.

> Unlike metaphor or allegory, which demand similar supplementing of meaning, irony has an evaluative edge and manages to provoke emotional responses in those who "get" it and those who don't, as well as in its targets and in what some people call its "victims."[1]

The kind of approach just described is not new but its position in the modern world is sometimes associated with the "postmodern condition"—the modern world may be suffering from a lack of faith in systems requiring extrinsic validation. To illustrate, where ethical standpoints were once central to artistic evaluation, a new kind of artifact emerged that did not submit to it. Art forms have increasingly appeared to distrust external criticism to the extent that they have sought to incorporate critical commentary within their own structures.[2] An earlier instance would be Jorge Luis Borges's mock critical essays about imagined writers and literature.[3] In cinema Quentin Tarantino's *Pulp Fiction* (1994) was a pioneering work that created a violent and amoral world through situations and characters but deliberately refused to take sides in the most brutal of occurrences, becoming more interested in the mechanics of the narration. Tarantino's film is an extreme case since it is difficult to say what the director's moral viewpoint is and this degree of ambivalence may not be found in the work of the filmmakers covered here. All of them are ethical in their thrust and have socio-political viewpoints but nonetheless try to conceal them with the expectation that their ambivalence will be seen through. It is significant that most of them have also produced more straightforwardly satirical works; in fact, their individual sensibilities could even be roughly described as "mocking." But this book is given specifically to those films *that are not openly satire* but more ambiguous and inviting interpretations. Irony depends upon interpretation; it is located in the difficult, unpredictable space between expression and understanding, which explains the disagreements over how something is meant, a likely outcome of the way readers will respond to the sections of this book.

This book is given to the satirical impulse in cinema and an important question is how satire and irony are related. The primary aspect is that irony is a rhetorical trope while satire is a genre. It is not easy to identify the actual signals constituting irony individually but they nonetheless exist and the feature important to satire is that its irony is at the expense of someone or something.[4] Satire and parody both use irony but the latter makes a comparison with another text through structural similarities/superimposition while the former uses irony only as a way of mocking the target without the mediation of another text. Parody and satire could also overlap when the text is used as a weapon. As an illustration, the sentiments expressed in the parodied text—like patriotic jingoism—could be a dominant political one in society at that particular moment in which case the text being mocked is not the ultimate target of the parody.

The satirical impulse

Satire, whether in film or literature, diminishes a subject and evokes, toward it, attitudes of amusement or scorn by using language that could be deliberately exaggerated. It is different from the comic in that while comedy evokes laughter as an end in itself, satire passes on a palpable attitude toward something, usually a recognizable entity. Since what is recognizable is shared with people in society, it is social satire that is most common. Satire is often looked down upon since mockery itself is not usually favored and represents its object not neutrally but as something deserving to be mocked. Comedy also mocks but what it mocks inhabits only the text while satire's mockery spills outside. But the impulse carries a disinclination to empathize, which can be valuable when empathy comes too readily. When empathy is demanded of any work of literature or film, it is arguably beneficial for it to be withheld to pass on understanding. To point out some characteristics of satire, it is topical; it claims to be realistic (although it is usually exaggerated or distorted); it is often shocking; it is informal in the sense that it deliberately avoids gravity; and (although often in a grotesque or painful manner) it is funny.[5]

Satire often took the shape of a monologue in classical literature as with the Romans in which the speaker/narrator is not quite the author but rather the author wearing a mask. Satire as monologue would be difficult in film but Sasha Baron Cohen's mockumentary *Borat: Cultural Learnings of America for Make Benefit Glorious Nation of Kazakhstan* (2006) can roughly be said to belong to the category or invoke it in some way. The director pretends that his film is a documentary about a Kazakh journalist in America and uses a series of naive interviews with unsuspecting Americans to lampoon (especially political) attitudes in the United States. A mockumentary is partly parody, which once again highlights the overlap between satire and parody.

Satire in classical literature is roughly seen to be composed of three categories: (1) the satirist, speaking either in his own person or behind a mask that is scarcely intended to hide, addresses us directly; (2) the satirist takes an existing work created with a serious purpose, makes the work, or the form, look ridiculous, by infusing it with incongruous ideas, or exaggerating its aesthetic devices; (3) the satirist relates stories in which he generally does not appear. Narrative, either as a story or as a drama, seems to be considered the most difficult type of satire in classical literature,[6] but it is also the kind that works best in cinema, which is realistic by nature since it is an extension of photography, catching what is before the camera.

If we compare realism and satire, both are interested in truths about the world, especially regrettable truths, but where realism does not find it necessary to mock them and may offer empathy as the neorealists did, satire mocks instead of empathize. Satire isolates the conditions or truths in order to chastise the humankind responsible for them. When realism grows too obdurate, it turns aggressive and becomes destructively bleak[7] since it does not allow laughter. If satire as a form is not favored, much of the controversy around it stems from its relationship to morality: Does the satirist have the right to attack society? Is his or her critique justified, and does it conform to a moral standard? Who is the satirist and who is satirized?[8] In each of the instances of satire cited earlier, the satirist puts on a disguise or speaks with another voice which can be said to be related to irony in which the author says something he or she does not mean.

This book is constructed around the belief that the ironical mode and satirical impulse are natural to certain kinds of authorial sensibilities. Since there is inevitably some kind of censorship in force on satire, it implies that such filmmakers could be driven into more complex kinds of expression by the restrictions placed on expression. Today, for instance, there is a degree of self-censorship operating and the pungent satire of some subjects—seen as deserving empathy—might be more difficult. By and large, satire is not regarded as compatible with ambiguity since what is mocked must be identifiable; but with such censorship, authorial sensibilities of a satirical bent need to reinvent themselves and try new forms of expression; but irony would then still be evidenced in the completed work. The filmmakers dealt with in this book are all complex in terms of their cinematic achievements and their satirical/ironic aspects need to be excavated through interpretive exercises. But they have one thing in common, which is that their sensibilities keep them away from empathy. Artists are not obliged to explain themselves and much of the satire/irony present in their work has often not been widely identified as such.

While the purpose of this book is interpretive—i.e. identify irony and satirical objective in filmmakers too complex to be termed "satirists"—we still need to first get a sense of what satire means and some of its affiliated categories. The term "ambiguity" is used in William Empson's first sense, ambiguity as metaphor, that is, when two things are said to be alike which have different properties. In many of the films dealt with in the book, something is apparently being said, which on deeper examination emerges as something different from what it is taken to be. Films evidently satirical are usually chosen to get a sense of the filmmaker's preferred targets or methods.

Satire and its affiliates

Satire has a number of associated categories and there is overlap between them, but if one were to proceed from the Romans, the three direct/explicit kinds are (1) Juvenalian (after the Roman satirist of the first century CE, (2) Horatian satire (after the Roman writer Horace, of roughly the same period), (3) Menippean satire (after the Greek cynic and polemicist Menippus of the third century BCE). There are also associated forms with a similar thrust like parody and burlesque. The earliest meanings of satire will not apply to cinema but certain common aspects can be inferred since there are common purposes for human expression regardless of age. It has been noted that Roman satire of all kinds was authentic predecessors of the novel,[9] which implies that we would be justified since cinema often draws from the novel.

Juvenalian satire

This is contemptuous and abrasive since Juvenal disagreed with the opinions of the public figures and institutions of the Republic and actively attacked them. A Juvenalian satirist is more likely to see the targets of his satire as evil or actively harmful to society and attack them with serious intent to harm their reputation or power. While Juvenalian satire often attacks individuals on a personal level, its most common objective is social criticism. The most famous satire of this kind from twentieth-century world literature would be George Orwell's *Nineteen Eighty-four* (1949) and *Animal Farm* (1945), Joseph Heller's *Catch 22* (1961), and Mikhail Bulgakov's *Heart of a Dog* (1925). In cinema, Stanley Kubrick's *Dr Strangelove or: How I Learned to Stop Worrying and Love the Bomb* (1964) would be Juvenalian satire as would *A Clockwork Orange* (1971). Elio Petrie's *Investigation of a Citizen above Suspicion* (1970) is a satire about a high-ranking police official who murders his mistress and then tests the strength of the police to charge him with the crime by deliberately planting clues that would lead to him in the normal course. It would be difficult in cinema to pick a real target and these films create fictional equivalents that would have historical or contemporary correspondence.

Horatian satire

A Horatian satirist makes fun of general human folly rather than undertake specific or personal attacks. In a work using Horatian satire, readers laugh at the characters in the story who are the subject of mockery, and society for behaving in such ways. It is satire in which the voice is indulgent, amused, and witty. Mark Twain's *Huckleberry Finn* (1884) and Nikolai Gogol's *Dead Souls* (1842) are examples. Fyodor Dostoyevsky's characters as those in *The Possessed* (1871–72) as well as those in Charles Dickens's novels (Wilkins Micawber in *David Copperfield*, 1850) could be out of Horatian satire. Jane Austen's Mr Collins and Lady Catherine de Bourgh (*Pride and Prejudice*, 1813) are satirical characters out of Horatian satire though most of the novel is not; in cinema, some of Eric Rohmer's early films like

Claire's Knee (1970), Robert Altman's *Nashville* (1975), as well as *M*A*S*H* (1970) seem to be, also. At the same time, satire is a recognizable genre while Rohmer's films and *Nashville* are too ambiguous to be recognizable as satire and they probably fall much short of being so. Sacha Guitry's films like *La Poison* (1951) can be more readily categorized as Horatian satire.

Menippean satire

This form is characterized by attacking mental attitudes rather than specific individuals or entities and tends to contain digressions, often on current and topical issues or pseudo-philosophical discussions of a comical nature. The philosophical discussions in Jean-Luc Godard's *La Chinoise* (1967) (e.g. with Francis Jeanson) could generally point to Menippean satire although they are too ambiguous to be called "satire." Fassbinder parodies the sequence in his first feature film *Love Is Colder than Death* (1969). The categories usually overlap and Gogol's *Dead Souls* is both Horatian and Menippean satire. Salman Rushdie's *Midnight's Children* (1981) and the novels of Milan Kundera (*The Unbearable Lightness of Being*, 1984) would also be illustrations. Lewis Carroll's *Alice's Adventures in Wonderland* (1865) has been considered an example as well. Sometimes what is satirical in a work of literature loses its satirical edge when transferred to film and Volker Schlondorff's *The Tin Drum* (1979) (partly Juvenalian satire) and Phillip Kaufman's *The Unbearable Lightness of Being* (1988) suffer because the tone of the narration is lost in the transfer. The tone of the first-person narrative of *The Tin Drum* is arguably compromised by the physicality of the presentation. An aspect of first-person literary narration is that the veracity of what is being said should be open to questioning, and that is lost when film enacts it before us, turning "word into flesh," as it were.

Burlesque and parody

Burlesque/parody has been succinctly defined as "an incongruous imitation"; that is, it imitates the manner (the form and style) or else the subject matter of a serious literary work or a literary genre in verse or in prose but makes the imitation amusing by a ridiculous disparity between the manner and the matter. The burlesque may be written for the sheer fun of it; usually, however, it is a form of satire since the texts comment upon the world and the mocking of a text is tantamount to mocking aspects of the world. Mel Brooks is a filmmaker who has specialized in parodies. Both *Young Frankenstein* (1974) and *Blazing Saddles* (1974) are parodies but *Blazing Saddles* is closer to satire since it deals with race issues. *Young Frankenstein* parodies a literary/film genre with fewer social echoes. Parodying a western is more subversive, politically, than parodying a horror or disaster film since the western was politically sacred at some point: it invoked a foundational mythology around the American state. Parody, in essence, is a form of imitation, but imitation characterized by ironic inversion not always at the expense of the parodied text.[10] The segment entitled "The Bonnie situation" in *Pulp Fiction* parodies a time management educational film without mocking it deliberately, and its relationship to what it is mimicking could also be termed pastiche.

"Teasing"

Satire includes a range of examples depending on its "bite" and if Juvenalian satire is at one end, teasing is on the other, the latter being a shallow parody of physical appearances. Where satire uses the comic against power and its oppressions, has a subversive character and a moral dimension that exercises judgment against its targets, teasing is a form of simple buffoonery, a form of comedy without satire's dissident edge. Teasing includes light parody, good-humored mockery, poking fun, and benign spoofs. Teasing typically consists of an impersonation of someone, monkeying around with his exterior attributes, physical blemishes, voice and mannerisms, quirks, way of dressing and walking, and/or the phrases he typically repeats. Teasing does not touch on the core issue, never makes a serious criticism judging the target with irony. It never harms the target's conduct, ideology, and position of power; it never undermines the perception of his/her morality and cultural dimension. Directed toward a powerful individual, it makes him/her appear more human and draws sympathy toward him/her. Mikhail Bakhtin observes that parody can be shallow or deep[11] and shallow parody is used to pay an author an indirect compliment. It is the opposite of "damning with faint praise" and designed to show that no further criticism can be made. *District 9* (2009) can, for instance, be read as a satire about apartheid but by having an alien category alongside the white and the black, the first representing the "truly oppressed," it subdues the horror of apartheid through the fictional aliens. *The Dictator* (2012) is another film that could be considered shallow satire. Fiction films dealing with democratic politics and institutions often tend to pull their punches and become shallow satire (Michael Ritchie's *The Candidate*, 1972).

The carnivalesque

This literary mode first enunciated by Mikhail Bakhtin[12] parallels the flouting of authority and temporary inversion of social hierarchies that, in many cultures, are permitted during a carnival. The literary work does so by introducing a mingling of voices from diverse social levels that are free to mock and subvert authority, flout social norms by ribaldry, and exhibit various ways of profaning what is ordinarily regarded as sacrosanct. Bakhtin traces the occurrence of the carnivalesque in ancient, medieval, and Renaissance writers (especially Rabelais). He also asserts that the mode recurs later especially in the play of irreverent, parodic, and subversive voices in the novels of Dostoevsky—novels that are carnivalesque. The famous scene in *The Possessed* (1871–72) in which the celebrated author Semyon Yegorovich Karmazinov reads his high-minded reflections to a gathering of village ruffians illustrates this.

The carnivalesque is not a commonplace phenomenon in cinema since authorial purpose undermines it; but as illustrations, one could cite the early sections of Volker Schlondorff's *The Tin Drum* or the Russian director Aleksei German's *Khrustalyov, My Car!* (1998). German's film is set in 1953, the year of Stalin's death, the communist state in disarray because the dictator is losing his grip and the repressive state structure is directionless. The political situation is grim but there is a deliberate cacophony in the visual detail, a refusal to focus solely on the dramatic elements. Federico Fellini's films could also be candidates for the carnivalesque if one got a sense of authority being mocked, but the absence of the political in films like *Amarcord* (1973) works against it. The multiplicity of voices from different social levels common to the carnivalesque is not a familiar experience in world cinema, where plot and teleology are necessary for a commercial appeal or the voice of *authorial narration* in auteur cinema (as in the case of Fellini's films especially) suppress the polyphony.

Sarcasm

Sarcasm in common parlance is sometimes used as an equivalent for all forms of irony, but it is more useful to restrict it only to the crude and taunting substitution of apparent praise for criticism and its obviousness presupposes an audience that is *already decided*; it does not need interpretation to detect it. It is sparingly used in major literature because of its crudity but in cinema its equivalent would be political propaganda, where the audience is already on the side of the filmmaker and need not be enlisted through persuasion. Propaganda films like those of the Nazis use it extensively.

All these categories just described use irony as a device; at the semantic level, irony is the marking of difference in meaning. It is brought about by the superimposition of semantic contexts, i.e. between and what is stated and what is intended; there are in effect two significations but only one signifier.[13] Since it is difficult to localize the irony textually, the evaluative function of irony—what it is being used at the expense of[14]—is often difficult to pinpoint. Salman Rushdie's novel *Midnight's Children*, which is usually termed "magical realist," Luis Buñuel's

The Exterminating Angel, and Eric Rohmer's *Claire's Knee* all exhibit it although the term satire does not quite fit them. We might propose that the ambiguity in much of the irony in the films dealt with in this book owes it to their ambivalence toward their subject material.

Enlisting empathy

The key tendency in the works described is to distance the reader/spectator from the action. But one of the characteristics of arthouse cinema today is arguably its tendency to elicit empathetic responses from its audience, identify with the protagonists to the extent possible. Why this is happening is difficult to identify but it does along with a general tendency limiting the pungency of what might be said about social situations, especially about selected human/social categories. Some of the things written by satirists like Jonathan Swift (e.g. *A Modest Proposal*) would be taboo.

One might suppose that given the egalitarian impulse informing much of the ethics in the modern age, empathy in cinema would have originated in the socialist worldview. Still, the history of cinema has "humanist" examples where the filmmaker takes a position close to the underdogs, trying to see with their eyes, but is not full of them. If one considers only the filmmakers preoccupied with social issues, the early filmmakers from the USSR, those like Sergei Eisenstein and Vsevolod Pudovkin, although they were Marxists, did not make films of this kind since they were primarily educating audiences to view things actively,[15] make them conscious of issues. In the Stalin era, with socialist realism being favored, the emphasis was on ideology, chiefly on finding historical analogs for the present[16] but the tone was usually heroic. Hollywood, around the same time, was producing film drama, the delivery of the story being primary, without an effort to make the spectator empathize with characters onscreen. Instances would be the films of Charles Chaplin (*Modern Times*, 1936), which although about the underdog are comic exercises and comedy effectively distances one from the subject and it would be difficult today to imagine similarly pessimistic comedies around the poor today. Parts of Chaplin's films may leave some teary-eyed but their attraction to the public at large did not rest on that. The aspects most valued today are arguably effective because of the comic distance Chaplin keeps from the drama. Overall, it would seem, cinema was not overly responding to social issues by drawing empathy through pathos.

The earliest films that actively enlisted the audience's empathy may have been those of the Italian neorealists. Neorealism, as is acknowledged, used children and old people as subjects, viewing them as victims of social ills, and one could cite the films of Roberto Rossellini (*Rome, Open City*, 1945, *Germany Year Zero*, 1948) and Vittorio De Sica (*Bicycle Thieves*, 1948, *Umberto D*, 1952). Children (by convention) are innocent, defenseless, and natural victims; they cannot be made responsible for the community at large. The neorealists did not stop with children but tended to reduce even adults to childlike states by conceiving of them simply, absolving them of responsibility for the society to which they belong. Critics of neo-realism have recognized that when adults slip back into innocence, they displace responsibility for the community elsewhere; they prefer to complain when the community fails to satisfy their needs and this complaining effectively replaces the possibility of action or critical thought.[17] Children have been used differently in cinema and an instance would be Luis Buñuel's *Los Olvidados* (1950) about slum children in Mexico City shown engaging in wanton cruelties.

Empathy is not always a laudable emotion to enlist since it promotes passive viewing. Bertolt Brecht's efforts were largely to undermine empathy in the Berliner Ensemble productions of plays like *Mother Courage and Her Children*. This is not a reiteration of the value of rationality since, as David Hume observed, what we call reason is actually a calm state of feeling and the most important thinking is imbued with feeling. Still, it is one thing to allow one's critical faculties to be active even while one is made to feel, and another to rely entirely upon arousing the emotions. Take, for instance, the dismissive charge that someone is arguing "emotionally"; if one believed in something, one would be emotional about it but still trust in reason, but if one did not, one would place more emphasis on the emotions, as demagogues do. The issue here is what emotions should be appealed to, and many artists have proceeded with the belief that distance is desirable since it would also facilitate reflection. Satire and its affiliates, as may be evident, stand at the other end in the emotions they try to enlist, away from empathy.

The chosen filmmakers and the approach

The promotion of empathy does not necessarily mean a decline in the satirical impulse—as much as a stricter selection of targets[18]—and cinema today does not lack for satire despite such a selection; there are obviously subjects especially considered satire-worthy.[19] The Coen brothers make films in which the satirical impulse is conspicuous even when the films are thrillers (*Fargo*, 1996) and there are others like the Greek filmmaker Yorgos Lanthimos (*Dogtooth*, 2009) and Radu Jude from Romania (*I Do Not Care If We Go Down in History as Barbarians,* 2018) who consciously choose satire. Ulrich Seidl (The *Paradise* trilogy, 2012–13) and Ruben Östlund (*Triangle of Sadness,* 2022) would also be more recent candidates who I have not selected. These are straightforward in their satire while irony is my primary interest. Satire does not need interpretation as in Östlund's film (about celebrities on a luxury cruise ship) and the filmmakers selected demand interpretation because of the ambiguity that marks their films.

Some of those I have selected—especially Buñuel, Fassbinder, and Kubrick—have been widely written about but their work is too complex for everything possible to have already been exhausted, and one still makes discoveries. If auteurs with irony and satire as their forte are the chosen field, it makes little sense to overlook them regardless of whether they have been written about before. But as a precautionary gesture, I am citing various other critical readings of the chosen films to demarcate my own position from the other critics. The Russian filmmaker Aleksei Balabanov is the least known among them but he carries forward the great tradition of Nikolai Gogol and Mikhail Bulgakov—in cinema—and I will try to demonstrate the importance of his work, argue for why he deserves a place at the very top alongside the others.

The eight individual filmmakers being studied in this book are not alike and there could even be disagreement that RW Fassbinder and David Lynch are satirical. But the fact is that all the filmmakers appear to have the kind of artistic sensibility that can be termed sardonic, producing at some point or other work with a satirical bite. Robert Altman made *M*A*S*H* (1970); Stanley Kubrick has made more than one satire while Paul Verhoeven's *Starship Troopers* (1997) has been seen to attack "fascist militarism." Fassbinder's *Katzelmacher* (1969), one of his most straightforwardly satirical films, is about the immigrant worker and the reception he gets in Germany. David Lynch may be a difficult filmmaker to pigeonhole but a part of *Mulholland Drive* (2001) can be arguably read as a satire on Hollywood, perhaps a stronger one than Robert Altman's *The Player* (1992). Luis Buñuel is usually termed a "surrealist"; surrealism—before it became mannered—was intended as mordant criticism of bourgeois conventions.

A factor bringing the filmmakers together is that where satire is usually transparent, these filmmakers have made films that can be described as puzzling, inviting interpretation but also resisting it. But there is no mistaking that the filmmakers are not prone to the enlisting of empathy from their audience and critical distance is what they hope for. Their bents are naturally satirical or ironic and are hence hard-pressed to hide it. Linda Hutcheon notes in her book *Irony's Edge* that ironic expression has become "outmoded" at the end of the twentieth century:

> My own particular interest was triggered by the fact that irony appears to have become a problematic mode of expression at the end of the twentieth century. It has never been without its problems, of course, but lately the various media seem to be reporting an increasing number of cases of the more or less disastrous misfiring of ironies. Magazine articles appear, exploring the prevalence of irony (Spy's March

> 1989 issue's cover featured Chevy Chase as "That Ironic Guy" and the heading, "Isn't it ironic?"). Or else they lament its rumored outmodedness (Esquire's September 1991 front page told us to "Forget Irony—Have a Nice Decade!"). And, certainly, if the newspaper and television coverage can be believed, today the public consequences of misunderstanding seem more serious, or at least more visible.[20]

Why irony became more "problematic" can only be conjectural since there was no official censorship in place but developments like political correctness may have influenced it; people needed to declare their political positions unequivocally. Many filmmakers covered in this book hence try contrarily to conceal their ironic or satirical intent and my interest is hence interpretive to find evidence of it. As already indicated, irony lies in the uncertain terrain between expression and interpretation and I have preferred the more ambivalent films to those that are satirical and do not invite interpretation. Where the tone in a film is often seen to indicate irony,[21] it may be difficult to identify it in many of the films dealt with in this book (as for instance Luis Buñuel's realistic films). But I will try to interpret the films based on their attitude toward their subjects in this way, as mocking or debunking something and identify that something. If there are restrictions on satire, it is the less obviously satirical films that invite interpretation.

Interpretation, as Susan Sontag proposed, may be the revenge of the intellect upon art[22] but, while the chapters in this book will be interpretive, they will also stop short of being reductive. The interpretations are those that a sensitive spectator rather than an academic affiliated with schools (like feminism or psychoanalysis) might respond to and the scrutiny pays particular attention to the actual texts. These directors are all those that the author has deep respect for and the effort will be to retain the mystery of their work and not simply to identify the satirical intent in it. Most of the directors have been prolific and each essay will focus only on a handful of their selected works. There will also not be a single approach toward all of them since they are markedly different and the same kind of detail with regard to one may not be relevant to another.

But the Afterword will also try to detect commonalities in their choice of targets and their strategies, which may be more broadly applicable across cinema. Their varied socio-political backgrounds are expected to increase the pertinence of this study. The overall effort is to make difficult films intelligible in some sense, with an eye to how audiences for whom the films were meant would likely understand them, although not to "explain" them threadbare. The emphasis is to speculate on each filmmaker's artistic vision through his work, most often in relation to the milieu he is from; biographical detail, unless especially pertinent to the interpretations, will not be dwelt upon. Since many readers may not have seen some of the films, an effort is being made to evoke them through descriptions alongside the analysis, which could also get them interested not only in the filmmakers but also in the ironic mode of expression. Each director is represented through a selection of films and the reason is that being complex in their visions, different films necessarily present different facets of each kind of expressivity.

NOTES

1. Linda Hutcheon, *Irony's Edge: The Theory and Politics of Irony* (London: Routledge, 1994), p. 2.
2. Linda Hutcheon, *A Theory of Parody: The Teachings of Twentieth Century Art Forms* (New York: Methuen, 1985), p. 1.
3. Borges's first work of fiction ("Pierre Menard, Author of the Quixote," 1939) used the strategy because the author had been gravely ill and feared that he might have lost his writing faculties. Trying fiction would be safer since he had not done that earlier and used his essay-writing skills to write it.
4. Linda Hutcheon, *A Theory of Parody: The Teachings of Twentieth Century Art Forms*, pp. 54–55.
5. Gilbert Highet, *The Anatomy of Satire* (Princeton: Princeton University Press, 1962), p. 5.
6. Highet, pp. 13–14.
7. Aaron Matz, *Satire in the Age of Realism* (Cambridge: Cambridge University Press, 2010), p. 2.
8. Johan Nilsson, *American Film Satire in the 1990s* (New York: Palgrave Macmillan, 2013), p. 1.
9. Mikhail Bakhtin (ed.), "Epic and Novel," in *The Dialogic Imagination: Four Essays*, trans. by Caryl Emerson and Michael Holquist (Austin: University of Texas Press, 1982), p. 22.
10. Linda Hutcheon, *A Theory of Parody: The Teachings of Twentieth Century Art Forms*, p. 6.
11. Gary Saul Morson, "Parody, History, and Metaparody," in *Rethinking Bakhtin: Extensions and Challenges*, ed. by Gary Saul Morson and Caryl Emerson (Evanston: Northwestern University Press, 1989), p. 73.
12. Mikhail Bakhtin, *Problems of Dostoyevsky's Poetics*, ed. and trans. by Caryl Emerson (Minneapolis: University of Minnesota Press, 1984), pp. 122–37.
13. Linda Hutcheon, *A Theory of Parody: The Teachings of Twentieth Century Art Forms*, p. 54.
14. "The second, evaluative, function of verbal irony has always been assumed but rarely discussed. Perhaps the difficulty in localizing irony textually has made theorists shy away from coming to terms with this other, but very important, function of irony—its pragmatic function. Most of them agree that the degree of ironic effect in a text is inversely proportionate to the number of overt signals needed to achieve that effect. But signals must perforce exist within the text in order to allow the decoder to infer the evaluative intent of the encoder. And irony is usually at someone's or something's expense. It would therefore

be in this pragmatic function, and not in the semantic one, that would lie the ready adaptability of mocking irony to the genre of satire." Linda Hutcheon, *A Theory of Parody: The Teachings of Twentieth Century Art Forms*, pp. 54–55.

15. MK Raghavendra (ed.), "A Trajectory of Form: The Development of Soviet/Russian Cinema (1910–2010)," in *Locating World Cinema: Interpretations of Film as Culture* (New Delhi: Bloomsbury, 2020), pp. 174–75.
16. Raghavendra, pp. 176–78.
17. Eric Rhode, "Why Neorealism Failed," *Sight and Sound*, 30.1 (Winter 1960), 27–32.
18. The sense that everything was fair game was exhibited by the satirical magazine *MAD*, which went into decline. The editor Al Feldstein recalled later: "We even used to rake the hippies over the coals. They were protesting the Vietnam War, but we took aspects of their culture and had fun with it. *Mad* was wide open. Bill (publisher William Gaines) loved it, and he was a capitalist Republican. I loved it, and I was a liberal Democrat. That went for the writers, too; they all had their own political leanings, and everybody had a voice. But the voices were mostly critical. It was social commentary, after all." Jason Heller, "Al Feldstein /TV / Interview," *The A.V. Club*, March 29, 2007.
19. This is arguably because of what "political correctness" has turned out to be. If one looks at the successful examples of twenty-first-century satire in cinema, one finds that the prominent ones are *Borat: Cultural Learnings of America for Make Benefit Glorious Nation of Kazakhstan* (2006) and Jordan Peele's *Get Out* (2017); they target anti-Semitism and racism in the United States, where there is a consensus on what the right attitudes are.
20. Linda Hutcheon, *Irony's Edge: The Theory and Politics of Irony*, p. 1.
21. A film's tone as its "attitude to its material and the stylistic register it employs" and irony is marked by distance, often signaled through several incongruities. First, incongruities are established between certain conventions of narration (for instance, the music) and the dramatic material narrated; second, and as a consequence, incongruities also emerge between the kinds of attitudes that would usually be adopted toward such conventions in their original context, and the attitudes adopted toward them here. James MacDowell, *Irony in Film* (London: Palgrave Macmillan, 2016), pp. 66–68.
22. Susan Sontag, "Against Interpretation," in *A Susan Sontag Reader* (Harmondsworth: Penguin, 1982), p. 98. But the kind of interpretation she had in mind was "deep interpretation" as in psychoanalytical approaches. The "depth" in a deep interpretation has little to do with profundity but, while surface interpretations presume that authors as agents are still in a privileged position with regard to what the representations are, deep interpretations presume that they have no such privilege. Surface interpretations, unlike deep interpretations, refer to the way a film's intended audience might understand a film. Arthur C Danto, "Deep Interpretation," in *The Philosophical Disenfranchisement of Art* (New York: Columbia University Press, 1986), pp. 51–53.

1

Luis Buñuel: Unaffiliated Radicalism

The term most frequently employed to label Luis Buñuel is "surrealist" and it would be necessary to make an association between "Surrealism" (the movement) and the expressed concerns of this book. Dada and Surrealism were terms originally used in art but surrealism means something much more now. Dada and Surrealism were "avant-garde" movements, the term "avant-garde," coming to signify the advanced socio-political as well as aesthetic position to which the modern artist should aspire. Broadly speaking, art in the nineteenth century was synonymous with bourgeois individualism. Owned by the bourgeoisie or shown in bourgeois institutions, it was a means by which members of that class could temporarily escape the material constraints and contradictions of everyday existence.

Modern art of the early twentieth century—the pictorial fragmentation of Picasso and Braque's Cubism, for instance—represented a startling break with traditional artistic conventions. Dada and Surrealism owed to Cubism and Expressionism, alongside Futurism, for their new pictorial languages, but they would have been uncomfortable with the idea, implicit in much of Cubism, that formal innovation alone provides a rationale for art. There was comparatively little stylistic homogeneity among the Surrealist artists involved, and literature was as important to them as visual art; if a common factor could be identified, it was confrontationist. Most notoriously, Marcel Duchamp submitted a men's urinal—playfully signed "R. Mutt" and titled *Fountain*—for the New York Society of Independent Artists exhibition in April 1917, thereby mocking the bourgeois idea of high art.[1]

The term "surreal" is misused so often that its association with the mocking impulse is hardly acknowledged today. The "surreal" assumes a kind of usage that advertising has taken advantage of: it often means "strange" in a way that invokes the paintings of Dalí or Magritte. Through the visual imagery of these two painters—more than any others of the Surrealist painters—surrealism has been transformed, for many people, into a category denoting a strangeness, or incongruity, brought about by juxtaposition.[2] Where surrealism was not a style and even satirist Jonathan Swift was "surrealist," if only in his rage,[3] surrealism is a moral sensibility and not concerned with the production of art. It is for this reason that it is difficult to name a representative surrealist in cinema after Luis Buñuel since most surrealists are only mannered. By and large, the word "surreal" now has such currency that it has marketable resonance.

This leads us to the "dream logic" deliberately conjured up by surrealism, what relationship that could have with satire. My proposition is that it is not so much the dream as the social/sexual repression the dream manifests that makes surrealism kin to satire. Dali and Buñuel's *Un Chien Andalou* (1929) has been interpreted psychoanalytically but can one forget the montage in it, the slicing of the eyeball compared to the sliver of a cloud across the moon? The juxtaposition contains the element of parody or satire where something taken as splendid is disparaged by comparison with the ugly or horrific. The concern of the second Dali/Buñuel film *L'Âge d'Or* (1930), which is more explicit, is the social repression of desire as an outcome of Catholic dogma.[4] Buñuel is usually interpreted psychoanalytically but psychoanalysis—which could once be startling—has been tamed as in academic study. Interpreting the films thus could even have an effect that the artists might have resisted—since Bunuel's target was the Church and repressed bourgeois existence that psychoanalysis could unsettle.

In his autobiography, *My Last Sigh*,[5] Luis Buñuel gives a whole chapter to his sister Conchita's memories, but immediately after a section dealing with the visit of the children to a cemetery, he inserts a memory of his own in brackets:

> After the war, I revisited this same cemetery, which seemed much smaller and older than I remembered. In the corner stood a small white coffin which had been pried open, exposing the mummified remains of a child, and a huge cluster of scarlet poppies that had grown up through what had once been its stomach.[6]

These are not Conchita's memories but Buñuel's and there is something especially surrealist about them, the juxtaposition of the gorgeous with the horrific, with a connection nonetheless being established. In the above image, for instance, the poppies have fed on the rotting carcass and scarlet is also the color of blood. But if Buñuel thought "surreally" he remained true to his creed and one still finds similarities between his first film (with Salvador Dali) *Un Chien Andalou* (1929) and his very last films, a work like *The Phantom of Liberty* (1974). But there are different phases, the first three films constituting one, his period in Mexico where he made commercially successful films being the second, and his last phase being when he was back in Europe, commencing with *Diary of a Chambermaid* (1964). The temptation would be to deal with each phase separately but it is more interesting to deal with two of his first three films as his "Surrealist phase," then with the Mexican melodramas (*El*, 1953), moving on to realistic films like *Diary of a Chambermaid* and concluding with his last surreal comedies.

The Surrealist phase

In his autobiography, Buñuel refers to the years 1929–33 as his Surrealist phase and the first film made by Buñuel (jointly with Dali) was *Un Chien Andalou* (1929). Describing it as "Surrealist" implies that it was made deliberately with an eye to the movement of that name and it would not be correct to describe all his works as Surrealist[7] in that sense. *Un Chien Andalou* was based on two dreams, one of Buñuel's where he saw a sliver of a cloud cutting across the moon like a razor slicing an eye and another of Dali's, a hand crawling with ants. When they made the film, the rule was that no idea or image that might lend itself to a rational explanation would be allowed. All doors had to be opened to the irrational, retaining only those images that surprised, without trying to explain why.[8]

There has been much writing on the Freudian implications of the film, not as research into psychology through outward observation but as psychoanalytic introspection, since the nebulous areas of the soul were among Freud's primary interests.[9] It was Buñuel who first directed critics to psychoanalysis as a means of investigating the symbols in the film,[10] and there could be a legitimate viewpoint in examining the film in the context of what sense was made of it at that time.[11] But Buñuel nonetheless went on to make films that demonstrated an attitude to society that was deeply oppositional and it is more pertinent from the viewpoint of this book to relook at the film today, what we can make of it, without this burden. Psychoanalysis today has perhaps become too associated with academia while to Buñuel it was still radical thinking. Buñuel's preoccupation throughout his career was debunking authority and I will only look at two of the three films from the Surrealist period here in order to be concise.

Un Chien Andalou (1929)

The title of the film means "an Andalusian dog" perhaps a reference to Buñuel himself imagining what his film's detractors might call him. The film has a woman protagonist and a male one but there is no intelligible storyline and the chronology is deliberately nonsensical ("once upon a time" followed by "eight years later," "around three in the morning," and "sixteen years ago"). There are separate scenes and continuity within them and the film is evidently narration of some sort; it is not, for instance, simply random images strung together and we follow the storyline even if it is deliberately unintelligible. By and large, it may be taken to revolve around unfulfilled sexual desire with the urge to inflict casual violence placed alongside. The notorious first scene is a metaphor and it is significant that what Buñuel dreamt of was the cloud slicing the moon; the comparison with the eyeball being sliced is his addition to the dream. The sequence with ants emerging out of the hand was Dali's dream and contains the kind of juxtaposition that was characteristic of his work and also of Buñuel's. A characteristic of the human hand is that it is particularly sensitive to touch and imagining it in contact with certain objects arouses repressed emotions. The image of the ants creates an intense feeling of discomfort, just as the man's hands on a woman's breasts connotes desire through a tactile representation.

There are other images that work in the same manner, as for instance the woman's underarm hair in the place of the man's mouth, and one could see an effort to excite repressed feelings through uncomfortable images. Underarm hair, for instance, is taken to be "indecent." Voyeuristic pleasure in casual violence is also taboo—the androgynous figure with the severed hand run over by an automobile watched excitedly by the protagonists. In contrast to these "irrational" images, the man pulling the two grand pianos upon which rotting dead donkeys are piled—and attached to Catholic priests—is actually devised to "mean," i.e. anti-clerical sentiment holding the Church culpable for repression since the man is unable to pull all this along even as he desires the woman. The last is perhaps too "intellectual" amidst a plethora of irrational images and seems contrary to Buñuel and Dali's desire to leave out whatever could be explained.

In his autobiography, Buñuel deals extensively with dreams and imagined scenarios difficult to explain but one immediately catches on to their irrational attraction. As instance is his account of his fight with Gala (later Dali's wife), throwing her to the ground and trying to choke her, but describing his intense desire at the time as only to see the tip of her tongue between her teeth. One might approach this through psychoanalysis but my own sense is that, in his early work, Buñuel was working with repressed urges, and film primarily as a way of exploring them. Many actions and gestures, that cannot be called anti-social, are prohibited on various grounds and *Un Chien Andalou* is an effort to give these impulses due representation as a way of acknowledging the irrational. "Civilization" and "culture" are held culpable for the repression but, if this is in the nature of a political statement, it is more an affirmation of expressivity rather than a demand for social freedom of action. When Buñuel thinks of a gag for a film, "a father shooting his own son because the boy dropped cigarette ashes on the floor,"[12] he is not proposing that as a legitimate act but as a taboo imagined. It is the hackneyed/religious sentiments associated with the parent-child relationship that he is confronting.[13]

Las Hurdes: Land without Bread (1933)

Before we go on to look at the surrealist documentary Buñuel made of a poor region of Spain, I should say a little about Buñuel's use of decoupage, and his juxtaposition of images. Normally film editing is equated with continuity editing in which the story is delivered or with "montage": montage is a term that largely describes the combining of two film elements, resulting in the production of a specific effect that could not be produced by either of the two elements.[14]

Montage assembles images to heighten the emotional impact of the action but decoupage cuts up the narrative into a procession of images, fragmenting the unified space to convey the story. For Buñuel in his Surrealist phase, decoupage was two or more successive shots independent of one another, autonomous, yet connected in some way or other. In order to obtain the maximum force of expression, cinematic composition would be an act of juxtaposition that causes readers to lose their geographical bearings.[15] It is pronounced in his Surrealist phase, where the effect derives from spatial contrasts. An illustration is in *Un Chien Andalou*, in which the man, seen in a medium shot, inquisitively gazes at his hand. His intense curiosity at what seems to be the sight of a swarm of ants crawling about a hole in his right hand (seen in extreme close-up). Another instance is the cutting of the woman's hairy underarm to a sea urchin. Dali, it may be noted, uses a similar kind of juxtaposition in his pictures, for instance a sensory organ—like an ear—to the expanse of the landscape and it can be seen as bringing irrational feeling and the outer world together. The ants on the hand are the picturing of a feeling (the hand going dead) as is the prickly feeling associated with the sea urchin.

Buñuel's later films are not visually striking; my own understanding is that since decoupage rather than montage was his emphasis, it was the events in the story that concerned him rather than the emotions "natural" to it—which is where those who emphasize montage score. It is the story (though unconventional) that produces the effects in *Un Chen Andalou* and rarely do we find visual composition and assembly getting importance in his more conventional films. Obviously, once his "Surrealist" phase ended, Buñuel could not use the visual juxtapositions of *Un Chien Andalou* but he used music, commentary, and imagery to produce the kind of effect he had in his first film with Dali.

Las Hurdes is ostensibly a documentary about a very poor area of Spain, which in the 1920s was not even connected by road. The film uses a flatly spoken commentary horrific in its impact. It begins with an annual ceremony in which riders ride under a rooster suspended upside down above the street and try to pull its head off. There are other such horrific representations: children drinking from the filthy trickle in which clothes are washed; villagers carrying topsoil in sacks from the mountains to make a rocky patch by a stream fertile with the information that if the stream floods (as it often does), a year's work is washed away. The area is prone to malaria and the voiceover tells us how to distinguish the Anopheles mosquito that carries malaria; we are also shown a villager shivering on the roadside from the disease. A child with sores in its mouth and throat is shown dying; a man bitten by a viper while clearing away some bushes to cultivate land has his hand infected; the area houses bees but the hives are owned by people outside; the hives are carried by donkeys and a donkey is shown killed by bees.

Among the most horrific images involve the "cretins" created by malnutrition and incest, wizened and violent creatures. The commentary also tells us that the people here have the same beliefs as anyone else and in the school shabby children learn that the three angles of a triangle are together equal to 180 degrees. We are also shown the relevant diagram in the textbook. One can see that the intended effect is contrary to the dry factual narrative maintained and the music by Brahms—i.e. one signifier and two different kinds of signified, pointing out irony.

The film speaks of horrific conditions but on examination one realizes that it is not an exposé that might lead to reform and much of it is staged. The donkey killed by bees could not have been caught accidentally and it is only the juxtapositions that lend the film such intensity. There is, for instance, a sequence in which a decrepit woman walks the streets at night announcing impending death. That is made known to us only by the commentary and the woman might have been anyone. The commentary is inexorably savage in its bluntness; liberal use of the word "cretin" is an instance. The cruelty of the words deliberately works against empathetic representations of poverty. The power of the film actually derives from its *non-elicitation of the expected emotions*. The area is impoverished and the only luxury visible is in the custody of the Church, with its habitual sanctimony. The music is also from a Brahms symphony, out of harmony with what is depicted. The film created a public outcry and it was later learned that Buñuel had even been earmarked for murder by the right-wing Phalange. It can be argued that the film acquires its unmatched power from the filmmaking and not out of the conditions actually observed. Rather than being a concerned picture of a social condition needing reform, the film is a surrealist exercise debunking the notion of civilization and human progress. The film is cruel but the Surrealists used "cruel" devices deliberately to disturb.

The Mexican melodramas

Buñuel's surrealist films were made as part of a movement, which he later left but he was looking for work desperately in the 1930s and, after working at dubbing films for Paramount in Paris and Warner Brothers in Madrid, he did directorial work on films for mass audiences for the film company Filmófono anonymously since they were not compatible with surrealism.[16] Of the eighteen films during his years at Filmófono, with him as executive producer, four are believed by critical consensus to have been directed by him, and two out of them were big commercial successes.[17] Buñuel understood film entertainment and could make films of all kinds. During the Spanish Civil War, Buñuel worked with the Republican government and the United States was engaged in making films about the Civil War. Buñuel had functioned as the coordinator of film propaganda for the Republic, which meant that he was in a position to examine all film shot in the country and decide what sequence could be developed and distributed abroad. The Spanish Ambassador suggested that Buñuel should visit Hollywood where he could tender technical advice on the films being made there about the Spanish Civil War.

Almost immediately upon his arrival in America however, the war ended and the Motion Picture Producers and Distributors Association discontinued making films on the Spanish conflict, and Buñuel could not return to Spain since the Fascists were in power. Buñuel did not succeed in the United States; Buñuel and his family "lived from one unsatisfactory crumb of work to another" because he "had none of the arrogance and pushiness essential for survival in Hollywood."[18] He was not flamboyant enough to capture the attention of Hollywood decision-makers, "In as much as [Buñuel], his wife and his little boy seemed to be such absolutely normal, solid persons, as totally un-Surrealist in the Dalí tradition as one could possibly imagine."[19]

It was in these circumstances that he was persuaded to work in Mexico. The Golden Age of Mexican cinema was peaking in the mid-to-late 1940s, and that was the time Buñuel was connecting with a Russian émigré producer named Oscar Dancigers who had been blacklisted in Hollywood on the charges of being a communist. Buñuel's first film, the musical *Gran Casino* (1947) was a commercial failure but he got a chance to make more films. *El Gran Calavera* (1949), a comedy, was however a smash hit and bought Buñuel the artistic freedom to make *Los Olvidados* (1950).

I will not deal with that masterpiece here since it does not come under the rubric of "Mexican melodrama" but, although about slum children in Mexico City, it is neither a document of poverty and delinquency nor a collection of objective details pertaining to the city's poor. Rather than being a sad-eyed gaze at disadvantaged children, the children in it are no more innocent than the adults, perhaps even less so. If the world is brutal, they take an active role in promoting that brutality, as in the scene where they rob a legless man, lifting him out of his begging cart and leaving him flailing on the sidewalk. But this film is too close to his later realistic work, and I will only deal with a few melodramas most interesting from my viewpoint.

Susana: The Devil and the Flesh (1951)

The film (made for a mass audience) begins with a young woman in a woman's reformatory being punished for bad behavior, confined to a separate cell with rats scurrying about. Those handling the violent young woman remark that she has been here so long but there is still no improvement in her conduct. The sequence is perfunctory and there is no possibility of our interpreting the woman as a victim of injustice. The woman then appeals to God to help her, tugs at the bars of the ventilator, which comes away in her hands: the miracle she asked for has happened. She climbs out in the pouring rain and escapes by crawling out under the barbed wire fence.

Susana (that is her name) finds shelter in the home of an upper-class landed family in which the people are the father Don Guadalupe, the son Alberto, who is just out of his teens, the mother Dona Carmen, the housekeeper Felisa and Jesus, the overseer in the estate; harmony evidently reigns. Don Guadalupe's favorite mare is giving birth to a foal but it is stillborn; the mare itself is in a fever. That is when Susana appears and is given shelter, fed, and clothed. The family is well-meaning; it only sees the beautiful girl as in need of succor and the protests of the housekeeper who sees her as dangerous are brushed aside. The workers however immediately see her as fair game, although perfunctorily and more as a diversion, leaving her alone thereafter.

The film is going against the grain—i.e. it is not portraying a beautiful woman among men as a natural victim and the working class as guileless; here the workers and the housekeeper immediately understand her but her upper-class protectors are fooled by her manner. Susana does the work assigned to her initially but dresses provocatively against Don Guadalupe's advice. Buñuel puts in an incisive anticipation of her likely effect on the household when Don Guadalupe kisses his wife fervently after meeting Susana and Dona Carmen asks why he is kissing her like that!

None of the three men show much interest in Susana but she works at making sure that they do, and the hacienda is literally torn apart by enmity: Jesus leaves, Alberto plans to move out of the house and live with her, and Don Guadalupe is also entangled. Jesus learns she is wanted by the police, does not reveal that but returns later for her, only to be rebuffed. Susana has subsisted on the strength of the secrecy that each man craves in his dealings, and matters come to a head when the conflict becomes so intense that everything hidden is now out in the open. The film ends with Jesus returning with the police to take her away. Jesus stays on as overseer and Don Guadalupe's favorite mare recovers as a way of connoting restored family happiness.

Susana is a melodrama in which a femme fatale enters the family domain and the formula has been widely written about. As Christine Gledhill notes,[20] melodrama is often centered on conflicts owing to the intrusion of a sexual character into the domestic spaces of the family. She proposes that melodrama's exposure of masculinity's contradictions poses a threat to the unity of the patriarchal narrative text. In most Hollywood melodramas involving the femme fatale intruding into the family (*Fatal Attraction*, 1987), the viewpoint is male and there is no doubt that the narrative is patriarchal. I would refer to *Fatal Attraction*'s sense of family togetherness with the man at the head, a rational man who is, in the final analysis, capable of making the right moral choices.

But Buñuel's film is different in that the men are helplessly in the grip of the seductress. The woman is a destructive force of the kind that Hollywood would perhaps not allow and her target is the bourgeois family that she virtually brings to ruin. I find the twist that the family has to be rescued by the police even savagely comic—in that it makes the sexual power exuded by the woman come into *competition with state authority*. In his Surrealist films, sexual repression in bourgeois society is a key theme and the same is present in a different avatar here. What *Susana* has been seen to stand for is the repression of instinct at the hands of bourgeois morals and traditional family order, and especially the weak network of norms and principles based on normative values.[21] Where the family enclosure is sacred but nonetheless an emblem of stability in Hollywood, Buñuel is emphasizing its fragility by having the state inadvertently become its protector—against the destructive power of feminine sexuality!

But this is not the cautionary portrayal of the threat to social stability from raw feminine sexuality. Feminine sexuality is rather like a force of nature, while the social fabric—created by the bourgeoisie—is so helpless against it that it needs its most powerful weapon—the "repressive apparatus of the state"[22]—to resist it effectively. Here again, Buñuel gets his effects by using irony against the emotional drama that one would anticipate from the unfolding story. It is also significant that the story proceeds from a miracle from within a religious institution. God's miracles, it would seem, are unmindful of the needs of a structured society.

El (1953)

El is another melodrama from Buñuel, this time about a husband's jealousy that verges on insanity. The film begins in Church where Francisco, a middle-aged aristocrat is participating piously in a ceremony. When the ceremony is over, he pursues a younger woman who he evidently knows from before, giving the slip to the Padre who holds him in high regard and wishes to introduce him to some dignitaries. Francisco lives alone in his ancestral home with a bevy of servants and he has a right-hand man Pablo who he trusts implicitly. When a maid complains of Pablo's sexual misconduct, Francisco asks Pablo to dismiss her. With that small detail, Buñuel establishes the patriarchal mores in Francisco's household that a wife would need to endure.

The woman Francisco is pursuing is named Gloria; she lives with her mother and is engaged to Francisco's architect friend Raul. When Francisco discovers this, he invites the couple to a party at his house and manipulates it so that Gloria marries him instead of Raul. The next scene is a flash-forward when Gloria runs into Raul and recounts her misery at Francisco's hands: he is incurably jealous and blames her conduct for his uncontrollable feelings. He is also deeply involved in a suit over some property that cannot be won and is obsessed with it. He wants Gloria to be nice to his attorney but is abusive and jealous when she does so. Gloria tries to leave him but he is genuinely contrite and her heart melts. Francisco, in accepted parlance, is only guilty of emotional cruelty but he has been extreme in that. He is so "normal" otherwise that not even Gloria's mother, who he has charmed, will believe her stories. The Padre, in fact, shoos her off when she tries to take his help since he knows all about Francisco through his confessions. Francisco, he tells her, had not even known another woman when he married.

In Francisco, Buñuel has created an authentic monster but although his portrayal might lend itself to psychological analysis, more important to the film from my perspective is the respect the monster commands even from Gloria and Raul. Francisco's descent into madness is inexorable and reaches its climax when he jealously follows "Gloria and Raul" into a church but finds his supposed quarry to be two strangers. This is followed by the laughter he thinks he hears from the congregation and he is so overcome that he assaults the Padre who he suspects of conspiring. The film concludes with Raul and Gloria now married and with their son "Francisco" visiting the monastery where the elder Francisco has been given a position as a novice. The disturbed man's conduct is reportedly exemplary but there are indications that he is far from cured.

There are several aspects of *El* that need commenting upon in light of what has already been argued. First, there is the motif of sexual repression among the bourgeoisie and its contribution to patriarchy. The entrance to Francisco's house is like the entrance of a castle, suggesting how forbidding it is to Gloria, shown to enter it more than once. An unused room dusted by Pedro gives us a sense of how ancient that patriarchy is and how much part of tradition. Second, the respect accorded to Francisco and Gloria's reluctance to leave him is indicative of a social structure in which the woman is naturally lower but is complaisant in it. The camaraderie between Francisco and Pablo, Pablo advising him to divorce Gloria and Francisco wondering if killing her might not be better underscores this. Pedro dissuades him with the brief remark that he would then have to spend the rest of his life in jail. There is no missing the fact that Pablo and Francisco are servant and master but they still collaborate against Gloria, Pablo only being Francisco's pragmatic side but being fully in tune with his feelings.

Then, there is the Church itself, entirely sympathetic to Francisco and where he is most comfortable. When the couple want entertainment, Francisco takes Gloria to the top of the Church steeple and expresses his hatred for the crowds below[23] emphasizing his superior position in society and his distance from the multitude.

Why Gloria and Raul should name their son Francisco is intriguing, but both of them evidently respect Francisco regardless of his despicable conduct. I would take this as indicative of the respect that hierarchy commands and the aristocratic Francisco is almost universally admired. Social respect, Buñuel is suggesting ironically, is not commanded by those with ethical principles (like egalitarianism) but those who follow the normative modes prescribed by institutions. Lastly, there is hesitation on Raul's part when the Padre asks him if the boy is his son. There is no indication that Gloria was pregnant when she left Francisco and we can take it as suspicion on Raul's part of the child's paternity, i.e. it suggests the likely *continuance of her status with her new husband* on account of his suspicions about her past. *El* is ostensibly melodrama but these elements make it a covert satire on all the things Buñuel habitually attacks: hierarchy, sexual repression, the bourgeoisie and the patriarchal mindset, and the collaboration of women in their own victimization. The property dispute also implies a parallel—Gloria as Francisco's treasured "property."

The realistic films

"Realism" is a term that has a large number of meanings but some of Luis Buñuel's films are realistic in a very specific kind of way and it will be useful to refer to Andre Bazin's writing here. In reality, or the real world, no event is ever endowed with an a priori meaning. The cinema's ontological purpose is therefore the reproduction of reality by respecting this essential characteristic. Cinema, in effect, must produce, or strive to produce, representations that are endowed with as much ambiguity as exists in reality itself.[24] Andre Bazin went on to draw lessons about montage from this perception since montage directs the gaze and imposes a meaning on the events in question, infusing them with purpose and significance that reality does not possess. I will use it with a different rationale, which is to emphasize that teleology is tantamount to a priori meaning, or the filmmaker playing God instead of respecting the ambiguity. My observation here is that any intervention by the filmmaker to provide a happy resolution to his or her story—as in the case of melodrama—would be tantamount to not respecting the ambiguity of the real world and trying to play God, as it were.

Before embarking on examinations of the two films dealt with in this section, *Diary of a Chambermaid* (1964) and *Tristana* (1970), I will briefly mention a "religious" film he made in Mexico, *Nazarin* (1959). In its celebration of the centennial of cinema in 1995, the Pontifical Council for Social Communications included *Nazarin* among the fifteen films in world cinema praised for their religious values,[25] which can only be pronounced a grave misunderstanding. In *Nazarin*, Buñuel presents us with a priest who has a Christ-like significance since he sets out to do good work and follows the image of Christ on the road followed by disciples. But his devoted followers are not men, as in the case of Christ, but two women, of whom one has a strong sexual drive while the other sells her body, and both, even if they are not completely aware of it, find the priest physically attractive, much to his unease.[26] Despite the protagonist's impeccable good intentions, he does more harm than good and that is Buñuel's way of demonstrating the ineffectuality of religion as a moral force. But my observation here is that it is essentially Buñuel's disinclination to play God by showing good intentions as fruitful (as would be comforting) that makes him a "realist" from a different mold. Buñuel's films are cruel because of their refusal to be comforting, their indifference to the poetic justice that most filmmakers cannot avoid in resolving their narratives. Two films that especially bring it out are *Diary of a Chambermaid*, previously filmed by Renoir in 1946, and *Tristana*.

Diary of a Chambermaid (1964)

To understand Buñuel's kind of realism, it would be helpful to compare his version with Renoir's also realistic portrayal. The films are based on the same novel written in 1900 as an expose of the landed bourgeoisie in rural Normandy. It tells the story of Celestine, a maid in the home of the Lanlaires, who have acquired their wealth through questionable means, and sees their quirks and perversions first-hand—which does not mean that it is sympathetic to the servant class. The worst person in the story is Joseph another domestic employee who is a murderer.

In Renoir's film, which was made in Hollywood, the family is ruled over by Madame Lanlaire while M. Lanlaire cannot assert himself. Their neighbor is Captain Mauger who has a running feud with the Lanlaires but develops an interest in Celestine, who he is even prepared to marry. The Lanlaires have a sickly son Georges who hates his family, lives in Paris but returns home briefly. His mother does her best to make him stay, enticing him through Celestine, buying clothes and finery to charm him. Joseph is intent on buying a café, marrying Celestine to help him run it, but he plans to pay for it by stealing the Lanlaire's silver. Celestine is torn between Captain Mauger, Georges, and Joseph; Joseph being a vicious scoundrel does not matter. Renoir is capable of portraying great cruelty (*La Chienne*, 1931) but after Joseph kills the elderly Captain who surprises him when being robbed, Joseph is killed and Celestine goes away with Georges. One senses that Renoir was under studio pressure in America to be optimistic, and the film's value lies in the portrayal of the amoral Celestine, who is happy to cooperate with Captain Mauger's murderer for entirely economic reasons. The initial sense of her nobility and goodness is quickly dispelled by her subsequent conduct. Still, the film lapses into an implausible conclusion by Celestine turning righteous.

Buñuel's version of *Diary of a Chambermaid*, made in France, is more unsparing than Renoir's. In it, Celestine (Jeanne Moreau) takes up an appointment with the Monteils; Madame is a frail person and Monsieur (Michel Piccoli) is a vigorous man with an insatiable sexual appetite that his wife cannot fulfill. Madame Monteil's grouse is that she has to pay for the pregnancies forced upon the female servants by M. Monteil's indiscretions. Joseph is the groom in this story, a brutal scoundrel though a devout Catholic and a "patriot," and Marianne the slow-witted scullery maid. Madame Monteil's father M. Rabour, has a shoe fetish and makes Celestine wear boots chosen from an assortment in his cupboard. Their neighbor is Captain Mauger, perpetually at war with Monteil, and flinging garbage into their garden. Claire is an orphan girl of ten who Celestine is fond of, and who eats with the servants. Both Joseph and Captain Mauger wish to marry Celestine, Joseph because he has his eyes set on a café that they could run successfully together.

One Saturday, M. Rabour is found dead clutching a pair of boots that Celestine has just worn. Claire, who has been hunting for snails in the woods, is missing and her raped body is discovered some days later; Joseph is responsible although no one knows or cares. Celestine, who was leaving, changes her mind and stays on when she learns of Claire, suspecting Joseph, who is eventually arrested though released for want of evidence. The film concludes with Celestine marrying Captain Mauger and Marianne working for them. The captain is attending to her wishes, as she lies in bed. Joseph has his café and a wife and is seen cheering a right-wing procession.

Buñuel's film is set around 1900 with the Dreyfus scandal in the air, and anti-Semitism the ruling sentiment. The first aspect of the film needing comment is its conveying of the sense of interpersonal disconnects. Everyone has his or her own private interests and none overrides any other. The Madame is different from the one in Renoir's film, not in charge except nominally. Her husband has no thought except with regard to his seduction of Celestine, which his wife does not hope to prevent. All she wishes for is that she should not be made to pay again for the pregnancy; M. Monteil is last seen seducing the slovenly Marianne inside a chicken coop when Celestine cannot be had.

Joseph in both films is evil, but there is a strange conviction to him here and a greater degree of complexity. After Claire's killing, Celestine offers herself to him as bait, wanting to frame him. But Joseph is not interested in "fun" but more in her as his wife and partner in his future café; her flirtatious conduct could make soldiers spend their money. When arrested after Celestine's machinations, he seems not even to care, since nothing can really come of it. He knows he has been framed but sees Celestine's deviousness as confirmation that "she is like him." Joseph's responses are contrary to what we might have expected of a rapist-murderer betrayed by his fiancée. Celestine called him a crook earlier and Joseph responded by citing his devoutness and patriotism. But at the bottom of this film is the sense (instantiated by both Joseph and a junior priest) that anyone associated with the military is socially beyond reproach. If there is no sense of a knit community which, despite the mutual hostilities, is palpable in Renoir's film, the only camaraderie exhibited here is between Joseph and the priest, who share the same political beliefs. The hatred of Jews and foreigners helps to keep it alive.

I described *Diary of a Chambermaid* as "realistic" in the sense that it is a godless universe that the film is describing and this is further evidenced by the seeming neutrality of its moral gaze. This is not in the sense that Buñuel has no moral viewpoint to offer or that his sympathies lie nowhere but that he adamantly refuses to play God. The rape and brutal murder of the 10-year-old Claire is horrific although it happens off-screen but the act does not enhance the taint on Joseph. If one were to describe the occurrence, Claire accosts Joseph returning through the woods on his cart, Joseph asks the girl to be careful because of the "wolves" about, smiles nicely to himself until an abrupt thought strikes him and he runs after her.

In another realist film, such an act would have repercussions upon the culprit, lead him into guilt or remorse or at least into a crisis of some sort, to a turning point as in Ingmar Bergman's *The Virgin Spring* (1960). Buñuel, of course, was not a believer like Bergman but there is something deeply disconcerting when the perpetrator of such an act in a film not only shows no evidence of being disturbed but also brushes accusations aside to embark upon career plans—fulfilled later. Claire's brutal killing has actually had no effect upon the perpetrator or on anyone else. Celestine acts against Joseph but is not affected when it is of no avail, which means that where her gaze might have represented a moral viewpoint, Buñuel resists the temptation of making it that.[27] If in his earlier films, Buñuel was against the Catholic Church, here he is against God; so complete is the sense in the film of no eye watching over humankind, alongside the devout being the worst. The irony of his gaze comes out of his deliberate subversion of expected outcomes like some retribution, remorse, or even regret seen. The connection between realism and satire in the introduction has perhaps not been manifested more strongly and the film might have become unbearably bleak without this irony.

Tristana (1970)

Tristana is not as highly regarded as some of Buñuel's other young-woman-seduced-by-old-man films but it illustrates the director's realist principles more acutely since it is not as sharp in its satire as, say, *Viridiana* (1961). In fact, one could even deny that its purport is satirical in any accepted sense. Still, it is deeply ironic by virtue of its deliberate eschewal of an anticipated teleology. To reiterate what has been said, Buñuel is much more concerned with narration than with obtaining emotional impacts in the accepted sense. His narration is "straightforward" but he inherently relies on *not following* an anticipated emotional trajectory, and intrigues us through it. The irony in his storytelling comes chiefly through this subversion.

In the film, Don Lope (Fernando Rey) becomes guardian to the impoverished orphan Tristana (Catherine Deneuve) and takes charge of her. Don Lope is an atheist and anti-clerical and appears to have views that Buñuel himself might have had but is also a libertine, although not without principles. In violation of one of these principles, he seduces his teenage ward and she becomes his mistress. Don Lope's cardinal sin seems to be to not take into account the possibility of the girl gaining maturity. Tristana has been acquiescent in her own seduction but grows to loathe her guardian. As luck will have it, she meets a handsome young artist Horacio, and leaves with him to become his wife. Don Lope is upset but he is also in straightened circumstances until his sister's sudden death brings him money. Two years later, Tristana returns with her husband, brought back because she is ill, suffering from a tumor in her leg. Horacio was willing to look after her but she wished to be restored to her former guardian. Horacio leaves and Tristana's leg is amputated; she now walks around on crutches or wearing a cumbersome wooden leg. In order to do the right thing, but with his sexual appetite not yet spent, the feebler Don Lope now marries Tristana. But she is another creature now, responding to his expectation of conjugality with summary laughter. Having to care for him, Tristana treats him with less care than he treated her, and coldly sees him to his death.

Tristana could seem a tale of feminine vengefulness but that must account for why Tristana leaves the husband she loves to return to Don Lope: is it entirely for revenge? There are also other aspects of the film without a clear purpose: Don Lope's housekeeper Saturna and her wayward deaf-mute son Saturno, who the crippled Tristana flashes her breasts at as a favor, and Tristana's abrupt loss of her leg. What the motifs mean and why Buñuel put them into the story need enquiring into since the film might have done without them—if revenge was its primary aspect. Saturna is capable and the contrasted waywardness of her son seems important to the film. Saturno cannot be reformed and a comparison is perhaps being made between Tristana and the boy, in that their natures are entirely implacable: they cannot be turned at all.

Tristana's leg is also an arbitrary development, of the kind that fiction tries to eliminate to make teleology intelligible. I propose that these secondary motifs are consciously introduced to loosen the causal link between Tristana's seduction and her subsequent hatred of Don Lope, the part she plays in his death. Don Lope is patriarchal but his ill-treatment by Tristana cannot be read as his feminism-inspired "punishment," but only on account of the loosened causation. Causation is a matter of providing explanations that translate into the director "knowing"; Buñuel, intent on not playing God, was conceivably disinclined to demonstrate such knowledge.

A visual motif of prominence in the film is Tristana's wooden leg which she rarely uses since it is painful, but it is seen lying around. This seems like a throwback to his Surrealist phase when fleeting objects/images were highlighted without a palpable purpose and we can ask the same question here: why introduce the leg when Tristana is hardly shown wearing it? Since the leg is usually lying on the bed, it could be argued that there is an association between the object and Don Lope's hopes of conjugality, which are persistently frustrated. The wooden leg could be taken to represent Tristana both as a metonym and metaphor. It is both part of her and a reminder of what she has become on her return to Don Lope, with no promise of life. But more pertinent than these interpretations is the fundamental narrative emptiness of the leg—since it has no rational function. But it serves constantly to tell us that the story is not following the anticipated teleology. If the leg had been "woven" into the narrative, it could not have had that effect.

Buñuel's realist films are taken to be a departure for him and there are indications that they were misunderstood by his own affiliates: "(Jean-Claude) Carrière has said that 'we pushed Buñuel to be Buñuel. Because [...] Buñuel's great defect was that he could sometimes be a realist. So, with (Serge) Silberman's approval, I had to push him as far as possible.'"[28] It has been argued in this section that Buñuel's realist films are not like the realist films of other directors such as Jean Renoir or the Italian neorealists because of his singular attitude toward creativity; he was averse to play God in the construction of his narratives because of his unbelief in Him. That gives his "realist" films a stark metaphysics not generally seen in examples of film realism.

The surreal comedies: Interpreting them

The set of films being written about in this section are difficult to describe but they are not realistic and cannot also be called fantasies although they deal with events that are not easily recognizable from life. But they draw laughter although one cannot be certain why, and that is why I am not inclined, for instance, to rely on psychoanalytical readings of *Belle De Jour* (1967) often comparing it to Michael Haneke's *The Piano Teacher* (2001).[29] There is an earnestness to psychoanalytical readings that suits Michael Haneke's film admirably but seems at loggerheads with Buñuel's work. Buñuel has not shown himself to be involved in any aspect of behavior other than its socio-political side and if anything stands out, it is his irreverence to all forms of pretension, including the scholarly kind. There may be objections to this on the grounds that even jokes have been interpreted psychoanalytically, but the psychoanalytical reading of a joke does not tell you why it is funny. The audiences for which a film like *Belle de Jour* is meant will find it comic[30] and that side should be important to any writer primarily interested in Buñuel as an artist. In the last film written about here, *Tristana*, there is no indication that it is the eponymous heroine's state of mind that Buñuel is exploring, or asking us to interpret. Her conduct, including the fact of her rage at Don Lope, is never asked *to be understood*.

I will not examine *Belle de Jour* here but try to demonstrate what kind of interpretation I am offering through a brief glance at another earlier surreal comedy *The Exterminating Angel* (1962). In this film—made in Mexico—a group of upper-class guests constituting the local bourgeoisie are unable to leave after a lavish dinner party, and they remain there for days although chaos has ensued after the servants have left. Here is Robert Stam on the film:

> Buñuel offers a more politicized version of the themes that obsessed the "theatre of theatre of absurd": entrapment (one thinks of Sartre's proto-absurdist *No Exit*); paralysis (Hamm in his wheelchair, his parents in their dustbin in Endgame); proliferating chaos (*Rhinoceros*); and the devaluation of language (*The Bald Soprano*). The film's central premise—the inexplicable entrapment of a pride of socialites—is as calculatedly implausible as those subtending many Beckett or Ionesco plays. The film also shares with absurdist theatre its affinity with burlesque film comedy. Beckett, like Buñuel, adored Buster Keaton, and collaborated with him on his own Film. Ionesco cited Groucho, Chico, and Harpo Marx as the three biggest influences on his work, and, in *The Chairs*, he has the old man impersonate the month of February by "scratching his head like Stan Laurel." *The Exterminating Angel*, like *The Chairs*, is structured on the comic formula of a slow descent from normality into anarchy, all performed in Keatonesque deadpan. The bear that wanders into *The Exterminating Angel* descends not only from the bear that frightens Pere Ubu in Jarry's play but also from the one that follows Chaplin along an icy precipice in *The Gold Rush*. Cannibalism is evoked in both films, and Buñuel's butler eats paper as if it were a delicacy, just as Chaplin dines on shoe and lace as if it were a gourmet meal.[31]

Stam tries to locate the film within a much larger cultural/literary context. But my interest would primarily be in the central plotting device or the "gag" of the dinner party itself, why the guests stay on despite extreme discomfort and nothing ostensibly forcing them to remain there for several days. Gags like Chaplin's in *The Gold Rush* are effective because they are invented on the basis of the essential human experiences inhabiting them and there must be a similar experience that Buñuel is depending on. The successful "gag" is what the audience has responded to and much of the interpretation in this book will depend on understanding that. The "gag" in Buñuel's film is not the descent into chaos but the fact that the guests are unable to leave; chaos could have been caused in any number of other ways. My interpretation of Buñuel's gag in *The Exterminating Angel* is based on one's experiences of exclusive upper-class parties where there are important people making contact; a guest is uneasy at leaving too early—perhaps for fear of being excluded from the circle and hence losing his/her own social consequence. In Buñuel's film personages from the Church, the military, the aristocracy, and business—pillars of the community who wield social power through their networks—all stay on after dinner and the only ones who leave are the servants, who are excluded anyway. Interpreting the central gags is an essential exercise in the kind of films I am dealing with in this book and I will now try to do that with two of Buñuel's last surreal comedies.

The Phantom of Liberty (1974)

This film may be the most difficult of Buñuel's films to describe although it is also one of the most thoroughly entertaining. The film does not have a plot but is composed of a large number of episodes strung together as if by chance. It begins in Toledo in 1808, with Napoleonic troops executing rebels, being a reference to the events behind Francisco Goya's painting *The Third of May 1808*, and then to necrophilia with the French officer in charge exhuming the body of a Spanish aristocrat Doña Elvira de Castañeda. But all this is only a story (by Gustavo Adolfo Bécquer) read in a park by a nanny while the little girl in her care plays. When playing, a stranger approaches the children and gifts them pictures they are not to show to adults. Still, that is what they do; the little girl's parents are duly shocked by the pictures but are nonetheless aroused into sexual feelings. The pictures, it turns out, are those of national monuments like the Arc de Triomphe and the Sacré-Cœur. The father has been an insomniac and, the same night, he sees himself visited in turn by a rooster, a lady with a candle, a postman, and an Emu pecking at the carpet. This induces him to visit his doctor the next morning but the doctor is not sympathetic despite his producing the letter from the previous night, delivered by the postman.

There are a host of hilarious situations in the film that one understands immediately but cannot quite explain: two bourgeois parents are told by their child's teacher that their daughter is missing; they visit the classroom and discuss the matter with the teacher even as their "missing" daughter draws attention to her own presence. Their response is to chide the little girl: she should not be disturbing adults when a matter of importance is being discussed. The parents then take the girl along to the police, to complain about her being missing; her presence is useful since an accurate description of her, without relying on hearsay, becomes possible. Months later the police commissioner informs the parents that the girl has been found. The father asks the commissioner if she was treated well and the mother confirms it through the commissioner. The parents depart with their daughter after duly thanking the commissioner.

The Phantom of Liberty follows its own logic consistently deliberately mixing up expected outcomes in situations:

> [S]yntactically the film has a perfect hyper-linearity as it always follows a narrative agent or object from one sequence to another, yet its misreading of semantic binaries creates ironic disjunctions and absurd ruptures—for example, socializing the wrong end of the digestive process (defecating rather than dining), or having the wrong outcome follow the guilty verdict at the murder trial of the sniper (dismissal rather than death), or searching for the lost little girl (because she is present rather than absent). We find ourselves concerned not only with structural choices (which character or object to follow into the next tale) but also with semantic decisions about the meaning of specific words (paraphernalia and pornography, guilt and innocence, freedom and tyranny). To explain these discursive challenges, Buñuel's Professor cites the relativism of cultural contexts and explicitly references Margaret Mead—an explanation also relevant to the disjunctive rhetoric of *Las Hurdes*.[32]

In the episode involving the professor above, he is there to lecture to policemen about the law but policemen are walking in and out of the classroom going out on assignments or coming in and the professor is not lecturing to the same policeman for more than a few minutes. The only ones present throughout are trying to pin the cut-out of a nude woman to his coat. But important here is that the professor is dealing with law as social convention relative to cultural contexts and (illustratively) posits a situation when he is invited to someone's home with his wife and those present sit around a table defecating in their respective potties but slip out to eat in secret. My own sense is that this segment contains the conceptual meaning of the film. David Bordwell noted that most films put in a line somewhere pointing to the conceptual meaning, as for instance "There's no place like home," is used in *The Wizard of Oz* (1939).[33]

Buñuel is drawing attention to the fact that most of our actions and expectations are guided by convention: children should avoid talking to strangers and the expected thing a suspicious stranger might gift an unattended child would be "dirty." We are so guided by conventions that our responses also follow them, although they might be inappropriate to the actual stimulus. In *Las Hurdes*, the children in the most backward of milieus are only taught that the three angles of a triangle add up to 180 degrees. In the episode involving the missing child, the parents are following two different conventions within the same story: the way parents are required to behave when their child is missing and the way they are expected to bring up their children—to be seen but not heard. In another episode, a sniper who has gunned down several persons in the marketplace is sentenced to death but then set free. There is a sense offered of judicial decisions being the outcomes of processes without the involvement of "justice." What is especially comic in the episode is that the killer wishes to sit down when the sentence is read out but is not allowed to, for its impropriety. *The Phantom of Liberty* is a surreal comedy but its socio-political import—so strong in his earlier films—is elusive. It would perhaps be helpful to relate its viewpoint with that of an earlier film with a strong socio-political slant so we are able to trace Buñuel's political evolution in the course of his career. I would suggest *Diary of a Chambermaid* and Joseph as a reference point to understand *The Phantom of Liberty*.

To recapitulate what was said earlier, after his impulsive rape and brutal murder of Claire, Joseph behaves without the slightest remorse and it is only being told about his involvement that clinches the matter, or we might have judged him as innocent. My argument is that there is an absence of underlying "continuity" in Joseph's behavior that points to basic urges and social convention being the causal agent in his acts rather than matters like conscience and moral choice. In essence, Buñuel proposes, we act in ways habitually dictated to us and one could cite the brutal conduct of normally peaceful men when they are required to behave as "soldiers" in battle. I would like to argue that, in *The Phantom of Liberty*, Buñuel has expanded on this notion of the absence of stable traits like conscience or moral choice lending spiritual "continuity" to the individual; social conventions ordain conduct more than these religious or moral issues. Phrased differently, we act as we are required to socially and not in ways that we should, based on an inner understanding of ourselves in the light of the moral instruction we have been made to receive. *El* also dealt with this although differently when the mad Francisco was respected for following conventions though he was so destructive.

Coming lastly to the strange ending of *The Phantom of Liberty*, two police commissioners (one of them under arrest for being an impostor) receive word that student radicals have assembled near the zoo, and they proceed there with a posse. One of the commissioners gives the other a cigarette (a gesture suggesting consensus)[34] and we are shown alarmed animals: a rhino, a seal, an elephant, a vulture, a cassowary, a lion, and an emu. The commissioner has asserted that a dead zebra is a small loss compared to the lives of policemen and we take that as a principle in the clash with the students, who are protesting "liberty." The last image is of a bewildered emu with no grasp of its own redundancy. I propose that the "innocence" of the animals of conventions acts as a counterpoint to the elaborate but futile structure constructed, piling convention upon convention through the string of episodes. Buñuel's earliest films attacked "civilized society" but it is only in *The Phantom of Liberty* that he gives us a clear sense of what that is.

That Obscure Object of Desire (1977)

Buñuel's last film tells the story of an upper-class man Mathieu (Fernando Rey) who feels an intense desire for a working-class woman Conchita (Carol Bouquet/Angela Molina), a chambermaid newly appointed. Mathieu immediately sets his heart upon Conchita and tries to seduce her, but unsuccessfully. The next morning, she is gone, but only for him to meet her accidentally three months later in Switzerland. Her friends hold him up in a park for a small sum, which she then tries to return. He learns that she lives in Paris with her mother and he follows her there.

There are two aspects to the film of great importance and the first is that the story is related by Mathieu in a railway compartment to his fellow travelers after they have just seen him emptying a bucket of water on a young woman, who is Conchita. The other is that Conchita is played by two women actors with different personas; why one actress plays her in some scenes and the second appears in others is not hinted at and neither do we get a sense of why. But Carol Bouquet has a much colder presence than Angela Molina, who is earthier but also capable of spitting fire.

The story related by Mathieu shows Conchita leading him along, never submitting to a seduction but always with a convincing reason. The first time in Mathieu's house it is since his dead wife slept in the same room, and then it is because she is a virgin and needs time to get accustomed to the idea. When he finds her mother in financial difficulties and tries to propose an arrangement, the reason is that he is trying to "buy" her, but the long and short of it is that, despite being physically intimate, she will not give herself to him. As this game goes on, Mathieu is more frustrated and disturbed and Conchita herself nastier until the moment comes when, after he has bought her a house, she declares her deep physical loathing for him and has sex with a young male friend before his eyes. But unlike in an Ingmar Bergman film (*Scenes from a Marriage*, 1973) that is not final since the next morning, she confesses that it was all play-acting and the young friend is gay; Conchita is still a virgin and only Mathieu will be allowed to have her.

One of the ways the film has been dealt with by academics is by dwelling on the notions of love and desire:

> How far are Mathieu and Conchita drawn to each other through what Ortega y Gasset refers to as the possessive nature of desire? How much—regardless of the choice of the word "désir" in Buñuel's title—through love? The obscurity in question may prompt reflection, through Ortega y Gasset, on the blurred meanings and overlaps between love

> and desire, but Lacan's notion of desire as a lack, something never satisfied—as opposed to a need or demand—provides an equally compelling perspective on the behaviour of a lover whose internal muddles are played out in his object of desire.[35]

My first difficulty with the approach is that this is the story as told by Mathieu, its primary purpose being his need to justify himself. While the story is undoubtedly corroborated to some extent since Conchita appears on the train and also drenches him with water, we cannot draw fine conclusions on the nature of their feelings for each other; to put it differently, hearsay cannot be conclusive evidence of them.[36] The fact that there are two actresses playing the role I would read as indication of his subjective sense of there being not one woman but two, given the contradictions in Conchita's conduct.

But there is another aspect here which is that Mathieu wanting to seduce Conchita who is his employee is never beyond reproach but he tells it anyway to his fellow travelers. When Conchita accuses him of trying to buy her from her mother, that is what he has been trying to do but Conchita refuses to exploit him, though financially in straightened circumstances. She even returns the key to the house he has bought her as a way of reiterating her love for him after her feigned sex with her friend. Apart from her vicious words to him, she cannot be blamed except not yielding to Mathieu but leading him on. I would take Mathieu's frank narration as an indication that he takes the listeners' sympathy as granted or, to phrase it differently, patriarchy as a given. The listeners (including a woman) are all from the bourgeoisie and whatever he believes will be shared. His sense of entitlement to Conchita will not be questioned by the others.

Mathieu's manservant Martin (reminiscent of Pablo in *El*) is an interesting presence here and he is also in sympathy with his master, going so far as to quote a philandering friend describing women as "sacks of excrement," implying the potency of both desire and gender conflict. Martin is of Conchita's class but is still in sympathy with Mathieu; even when Mathieu has hit Conchita: he "understands." Mathieu and Conchita are together at the end but that is only underscoring the stability of a symbiotic relationship founded in reciprocal abuse but with its enduring basis in sexual desire, which imposes interpersonal tolerance. What Buñuel is doing is presenting a picture of the relationship from a patriarchal viewpoint with the expectation that we will catch on to the subjectivity of the portrayal; the use of the preposition "that" in the title corroborates it since it implies pointing out something from an assumed position.

The "love story" of Mathieu and Conchita happens in the foreground and an issue needing examination is the terrorist violence in the background, notably from the Revolutionary Army of the Infant Jesus (RAOIJ). Buñuel had once joined the Communist Party when in Spain but he was perhaps too independent in his thinking to be drawn to any ideology; RAOIJ is evidently an indication of the arbitrariness of radical causes, which do not guarantee even ethical concerns; the association of the term "revolutionary" with Christianity supports this. His two last films both include segments implying political unrest and so-called radical movements underneath but both films are also about the stability of structures impervious to radicalism. Structure in *The Phantom of Liberty* was represented by bourgeois society and its conventions while in *That Obscure Object of Desire* it is patriarchy that represents it. These last comedies (including *Belle de Jour*) testify to the same preoccupations Buñuel had shown at the commencement of his career as an implacable surrealist: sexuality, bourgeois convention, patriarchy, class divisions, and religious repression being the primary ones. The last two films, though comedies, are as pessimistic as radical visions in cinema have ever been. At the end of his career, laughter apparently became to Luis Buñuel an expression of political hopelessness.

NOTES

1. David Hopkins, *Dada and Surrealism: A Very Short Introduction* (Oxford: Oxford University Press, 2004), pp. 2–9.
2. Michael Richardson, *Surrealism and Cinema* (New York: Berg, 2006), p. 2.
3. Jonathan Swift, "The Platform of Prague Twenty Years On," in *Surrealism against the Current: Tracts and Declarations*, ed. by Krzysztof Fijalkowski and Michael Richardson (London: Pluto Press, 2001), p. 82.
4. Linda Williams, *Figures of Desire: A Theory and Analysis of Surrealist Film* (Berkeley: University of California Press, 1981), p. xi.
5. Luis Buñuel, *My Last Sigh*, trans. by Abigail Israel (New York: Vintage, 1983).
6. Buñuel, p. 44.
7. Judging from his writing, irrational expression bordering on scandal was at the basis of surrealism, with an intense desire to break with social convention, "The surrealists answered every question with what seemed to be total frankness—a feat that might seem commonplace today, but at that time, questions on the order of 'What's your favourite place to make love? With whom? How do you masturbate?' seemed incredible to me." Buñuel, p. 105.
8. Buñuel, p. 107.
9. Bruno Bettelheim, *Freud and Man's Soul* (London: Fontana, 1985), p. 68.
10. Frank Stauffacher, *Art in Cinema* (San Francisco: San Francisco Museum of Art, 1947), pp. 29–30.
11. Sarah Cooper, "Surreal Souls: *Un Chien Andalou* and Early French Film Theory," in *A Companion to Luis Buñuel*, ed. by Rob Stone and Julián Daniel Gutiérrez-Albilla (Chichester: Wiley-Blackwell, 2013), pp. 141–55.
12. Buñuel, p. 120.
13. Scandal was what the surrealists were primarily interested in creating and this had an effect upon their lives. "When I arrived in Figueras, it was Christmas 1929. At first,

all I could hear were angry shouts; then suddenly the door flew open and, purple with rage, Dali's father threw his son out, calling him every name in the book. Dali screamed back while his father pointed to him and swore that he hoped never to see that pig in his house again. The rage of the elder Dali was perfectly understandable; it seems that on one of his paintings then at a show in Barcelona, Dali had written—in black ink and large, sloppy letters—'I spit joyfully on my mother's portrait'!" Buñuel, p. 121.

14. Jacques Aumont, Alain Bergala, Michel Marie, and Marc Vernet, *Aesthetics of Film*, trans. by Richard Neupert (Austin: University of Texas Press, 1992), p. 47.
15. Tom Conley, "Buñuel Entomographer: From Las Hurdes to Robinson Crusoe," in *A Companion to Luis Buñuel*, ed. by Rob Stone and Julián Daniel Gutiérrez-Albilla (Chichester: Wiley-Blackwell, 2013), pp. 188–89.
16. Manuel Rotellar, "Luis Buñuel en Filmófono," *Cinema 2002*, 37 (1978), pp. 36–40.
17. Roger Mortimore, "Buñuel, Sáenz de Heredia, and Filmófono," *Sight and Sound*, 44 (Summer 1975), pp. 180–82.
18. Ruth Brandon, *Surreal Lives: The Surrealists 1917–1945* (London: Macmillan, 1999), p. 358.
19. John Baxter, *Buñuel* (New York: Carroll & Graf, 1994), p. 172.
20. Christine Gledhill, "The Melodramatic Field: An Investigation," in *Home Is Where the Heart Is: Studies in Melodrama and the Woman's Film*, ed. by Christine Gledhill (London: British Film Institute, 1987), p. 10.
21. María Pilar Rodríguez, "*Susana*: Melodrama and the Voluptuosity of Destruction," in *A Companion to Luis Buñuel*, ed. by Rob Stone and Julián Daniel Gutiérrez-Albilla, p. 242.
22. The term used by Louis Althusser to characterize state institutions like the police and the military. Louis Althusser, *Lenin and Philosophy and Other Essays*, trans. by Ben Brewster (New York: Monthly Review Press, 1971), p. 137.
23. Alfred Hitchcock's *Vertigo* uses similar imagery in the same sinister way and Hitchcock perhaps saw *EL* at some time.
24. Jacques Aumont, Alain Bergal, Michel Marie, and Marc Vernet, *Aesthetics of Film*, trans. by Richard Neupert (Austin: University of Texas Press, 1992), p. 54.
25. Maria Elena De Las Carreras Kuntz, "Luis Bunuel's Quarrel with the Church," *Crisis Magazine*, November 1, 1999. https://www.crisismagazine.com/1999/luis-bunuels-quarrel-with-the-church. Accessed May 19, 2020.
26. Gwynne Edwards, *A Companion to Luis Buñuel* (Woodbridge: Tamesis, 2005), p. 131.
27. Buñuel's Celestine is a far cry from the innocent girl from the novel by Octave Mirbeau: "His Célestine, an impenetrable, calculating, and ultimately powerful character who sexually manipulates men is a far cry from her Mirbellian ancestor who writes in her diary: 'They [men] only have to begin talking to me, the monsters, and directly I feel the warmth of their breath and the pricking of their beard on the back of my neck, it's no use. I just go as limp as a rag, and they can do what they like with me.'" Kate Griffiths, "Spectral Cinema:

Le Journal d'une femme de chamber," in *A Companion to Luis Buñuel*, ed. by Rob Stone and Julián Daniel Gutiérrez-Albilla, p. 401.

28. Julie Jones, "Luis Buñuel and the Politics of Self-Presentation," in *A Companion to Luis Buñuel*, ed. by Rob Stone and Julián Daniel Gutiérrez-Albilla, p. 93.
29. Jimmy Hay, "Inside/Outside: Space and Sexual Behavior in *Belle de jour* and *La Pianiste*," in *A Companion to Luis Buñuel*, ed. by Rob Stone and Julián Daniel Gutiérrez-Albilla, pp. 554–70.
30. Taking only the opening scene of the film, it is evident that Buñuel intended to draw laughter. Severine and her husband are on a coach driving through a forest and there is a minor disagreement, with Severine remaining stubborn. The husband responds by having the coachmen pull her out, tie her to a tree, strip her of her upper-attire, and flog her, after which her husband permits the coachmen to do "what they will." A coachman then sets about kissing her bare back with relish after which they set about to flog her. But all this is only a daydream since it cuts to Severine in a reverie, her caring upper-class husband asking about her thoughts.
31. Robert Stam, *Reflexivity in Film and Literature: From Don Quixote to Jean-Luc Godard* (New York: Columbia University Press, 1992), pp. 185–6.
32. Marsha Kinder, "The Road and the Room: Narrative Drive in the Films of Luis Buñuel," in *A Companion to Luis Buñuel*, ed. by Rob Stone and Julián Daniel Gutiérrez-Albilla, p. 448.
33. David Bordwell, *Inference and Rhetoric in the Interpretation of Cinema* (Cambridge, MA: Harvard University Press, 1989), pp. 8–9.
34. Sharing of cigarettes by society's leaders becomes a way of affirming that a proposed action follows laid down convention; it puts the seal of social approval on something and is the sense to be obtained.
35. Peter William Evans, "Splitting Doubles: Ángela Molina and the Art of Screen Acting in Cet obscur objet du désir," in *A Companion to Luis Buñuel*, ed. by Rob Stone and Julián Daniel Gutiérrez-Albilla, pp. 494–95.
36. In writing on Hollywood, David Bordwell notes that flashbacks are motivated by character memory but they are not restricted to point of view. Extended flashback sequences include material that the remembering character could not have witnessed or known. David Bordwell, "The Classical Hollywood Style," 1917–60, in *The Classical Hollywood Cinema: Film Style and Mode of Production to 1960*, ed. by David Bordwell, Janet Staiger, and Kristin Thompson (London: Routledge, 1988), pp. 42–43. In *That Obscure Object of Desire*, however, the narration constantly goes back to the telling of the story by Mathieu. There is also nothing in the film that corresponds to Conchita's viewpoint. In effect, there is nothing shown "as it happened" except in the framing sequence with the train passengers and after when Mathieu have left the train together.

2

Stanley Kubrick: The Failure of Human Systems

Writing about an American filmmaker usually involves a different approach from writing on European arthouse cinema—in that American cinema is usually attached to genres, and most of the best American films are generic. The style favored in classical cinema was also an "invisible style" where the principal purpose was the delivery of the story unhindered. Directors could be associated with genres like musicals (Vincente Minelli), noir (Nicholas Ray), westerns (John Ford), comedies (Billy Wilder), or they could work in a variety of genres (Howard Hawks). Their relationship with a genre could also be less straightforward and could be in the nature of a revision of conventions (as Sam Peckinpah with the western). Later, under the influence of European art cinema, American directors also later made films different from the earlier generic exercises and examples would be Francis Ford Coppola's *The Conversation* (1974) and Robert Altman's *Three Women* (1977), which are ambiguous in deliberate way that American cinema was not known to be.[1] Newer directors of the postmodern age are Quentin Tarantino (*Pulp Fiction*, 1994) and David Lynch (*Blue Velvet*, 1986), the latter being the more difficult of the two to characterize. They too work with genres but more as pastiche or parody.

Stanley Kubrick is different in that his films are not easy to associate with genres and are more likely to be based on novels of short stories though he makes changes. After making two noir films, his literary sources have been varied, ranging from Vladimir Nabokov (*Lolita*, 1962) to Stephen King (*The Shining*, 1980). He began as a photographer and his visual sense is impeccable; the film influences he cites are therefore formal like Max Ophuls (*The Earrings of Madame de*, 1953) whose tracking camera he tried to emulate. He is seen as a perfectionist and took time planning each film. He is difficult to categorize in terms of thematic concerns but his vision is broadly "ironical" although he may have made only two straightforward satires.

As with the other directors, my proposition is that being "ironical" is a quality in the filmmaker-auteur's visions that they cannot easily rid themselves of and I will try to demonstrate this in Kubrick.[2] Someone prone to being ironical may be drawn to satire and Kubrick's films are evidence. But I will examine only a few, grouping them into two categories and beginning with three anti-war films: *Dr Strangelove, or How I Stopped Worrying and Learned to Love the Bomb* (1964), *Paths of Glory* (1957), and *Full Metal Jacket* (1987). *Dr Strangelove* is explicitly satirical and will be considered first, and then the other two to demonstrate that irony comes most readily to Kubrick even when not producing obvious satire. War and humankind's capacity for self-destruction is a key notion in Kubrick's films, which is why the first section looks at his anti-war films. The remaining films chosen, not widely regarded as satirical, I examine in chronological order: *2001: A Space Odyssey* (1968), *The Shining* (1980), and *Eyes Wide Shut* (1999). I am not including *A Clockwork Orange* (1971), which is too obviously satirical and my primary interest is in irony.

The anti-war films

The first film in the category is not "anti-war" like the other two, although impending war is its subject. Though broadly futuristic, *Dr Strangelove* does not specify a future time and could be in the present. Kubrick's films, by and large, are not "difficult" as, say, European art filmmakers are and yield their meaning immediately and this is true of *Dr Strangelove*, perhaps Juvenalian satire since it attacks folly and social/political conditions viciously. This means that academic responses also do not rely on theories (like psychoanalysis as in the case of Buñuel) but look at him as a filmmaker with a socio-political viewpoint to offer and a filmmaking strategy. But since Kubrick's works are so varied in terms of the subjects, they take up and my purpose is to find unity within his oeuvre, I will endeavor to look beneath what he is saying to study deeper attitudes toward the world and human activity than may be visible.

Dr Strangelove (1964)

The film opens in Burpelson Airforce Base headed by General Jack D Ripper (Sterling Hayden) which houses the Strategic Air Command (SAC) 843rd Bomb Wing, flying B-52 bombers armed with hydrogen bombs. The 843rd Wing is flying on airborne alert two hours from their targets inside the USSR. General Ripper orders his executive officer Group Captain Lionel Mandrake (Peter Sellers) of the UK Royal Air Force, to put the base on alert, and to issue "Wing Attack Plan R" to the patrolling aircraft, an order to bomb targets inside the Soviet Union. The film beginning with this, it is evidently the first cause and General Ripper is acting on his own initiative.

Dr Strangelove is a comedy, a black one to be sure, and its target is the elaborate "defense" system put up by the two superpowers which they have no means of handling right. The film takes place at three major levels: the Airforce Base with General Ripper and Mandrake exchanging words until Mandrake realizes that the order issued is completely without rationale but it is also too late to recall the B-52s. The second level is the War Room at the Pentagon where President Merkin Muffley (Peter Sellers) is meeting senior officers among whom General Buck Turdgidson (George C Scott) is the USAF Chief-of-Staff. Also present in the War Room is Dr Strangelove (Peter Sellers), a scientific advisor of German origin and once an advisor to Adolph Hitler. Called in is the Soviet Ambassador Alexi de Sadeski who establishes contact between the President and Soviet Premier Dimitri Kissov. The third level is the B52 commanded by Major TJ "King" Kong (Slim Pickens) representing the squadron of strategic bombers.

Communications are down and the plot revolves around the bombers having to be recalled but this not being accomplished since General Ripper has cut off all communications. General Ripper believes that the Soviets are polluting the precious bodily fluids of Americans through "fluoridation" of American water supplies. The President authorizes an attack on the Airforce base and General Ripper kills himself without revealing the recall code word. Mandrake guesses it and makes a call to the President from a public phone, for which he has too few nickels. Col. Bat Guano (Keenan Wynn) who helps him with nickels gets it from a Coke vending machine, warning Mandrake that he will be answerable to the Coca-Cola Company.

In the War Room, Premier Kissov reveals that if the USSR is bombed, it will set off the Doomsday Machine, which will envelop the earth in radioactive clouds for 93 years. Why would anyone invent a Doomsday machine? The USSR was worried about a "doomsday gap" developing with the United States, learned from the *New York Times*. And why did they keep it secret since its value as a deterrent depended on it being widely known? The Premier was planning to announce it at the next Party Congress due within a few days. At every stage, Dr Strangelove (in a wheelchair) is consulted, a man who has problems keeping his right hand down since it shoots up to make a Nazi salute on its own, out of habit. Dr Strangelove lets the President know that the only way out of total annihilation would be for chosen Americans to live in mineshafts suitably appointed, ten attractive women to every man, to procreate and produce a new Master Race. The man-woman ratio has the lascivious General Turgidson interested. There is, however, the possibility of a "mineshaft gap" developing with the Soviets to be guarded against.

The strategic bombers are eventually recalled with active cooperation between the President and Premier Kissov, that is, all except the one commanded by Major "King" Kong, which though damaged, resolutely pushes though and drops a bomb on a backup target, setting off the Doomsday Machine, with the World War II song "We'll Meet Again" heard on the soundtrack. Major Kong in a cowboy hat is last seen riding the bomb to its destination on the ground. He was sitting on the bomb to set right an electrical problem and the bomb hatch opened on its own.

Dr Strangelove is a hilarious take off on the issue of nuclear deterrence that became particularly relevant after the Cuban Crisis in 1962. But an aspect of the film deserves comment and that pertains to the viewpoint taken by Kubrick. I would draw attention to the bomber commanded by Major Kong making its way to the target despite being damaged. Normally, when suspense is built up like this in a film, something damaged made functional against all odds, the suspense works toward affirming a desirable goal and its achievement. In this case, the goal is setting off humanity's destruction, which is evidently not desirable. My observation is that members of the audience are led to wish for failure rather than success in Major Kong's endeavor. Regardless of the "satire," the natural human urge would be the avoidance of destruction but the film is heading there by trying to promote the opposite anticipation. The ending where the whole world goes up in a series of explosions to the sound of optimistic music also has us wondering: where does Kubrick take his position in this story?

Satire of *Dr Strangelove*'s kind is dystopian but an imagined dystopia distinguishes a desirable situation from an undesirable one. Taking George Orwell's *Animal Farm* (1945) or Joseph Heller's *Catch 22* (1961) as examples of Juvenalian satire, regardless of the mockery that permeates them, there is little doubt that the novels posit wished-for situations even when these do not actually materialize. In *Animal Farm*, it would be the farm animals' efforts at self-government becoming successful, and in *Catch 22*, despite all the horrors it deals with, it would be Yossarian escaping from the War. *Dr Strangelove* is different in that Kubrick, by imagining and depicting humanity's destruction facetiously, is taking a position outside it, as it were. It is as though Kubrick was not of this world, or disavowing it—essentially as not worth saving. There is a reading in the film that the destruction of the world by the Doomsday Machine is only the penultimate ending. It is only the means toward an ultimate "Final Solution"—to clear the way for a Master Race selected by Dr Strangelove to survive underground.[3] Even if one accepts this reading, it still gleefully presupposes the end of the world as we know it.

Kubrick, like Swift, apparently believed that: "A satirist is someone who has a very skeptical, pessimistic view of human nature, but who still has the optimism to make some sort of joke out of it. However brutal the joke might be,"[4] but this remark needs to be interrogated. Jonathan Swift could genuinely shock with writing like *A Modest Proposal* (1729) in which he suggested that poor Irish people could sell their children to rich English gentlemen and ladies, to be eaten. The reason why this cannot be applied to *Dr Strangelove* is that Kubrick's film hardly shocks like Swift's outrageous proposal since it is too close to reality; the world was on the verge of nuclear war in 1962. To shock, even the imagining of what is said should be liable for proscription. The "joke" he makes about the destruction of humankind is not tinged by horror. The reason, it would seem, is Kubrick not taking his position among humankind but adopting a lofty perspective. The use of "We'll Meet Again" on the soundtrack to the destruction of the world is also problematic since it underscores this lofty perspective of someone untouched; the music not being diegetic, it represents an authorial imposition on the narrative.

Before looking at the two remaining anti-war films, it would be helpful to distinguish between what seem to be two separate generic categories—the war film and the anti-war film. The war film could be pacifist like Terrence Mallick's *The Thin Red Line* (1998) but, if it affirms welcome aspects of war like camaraderie, and shows war as a given in the process of making that affirmation, it would still not be "anti-war." In Mallick's film, a former deserter Private Witt dies heroically when the situation so requires him to. Certain wars evidently lend themselves to "anti-war" statements while others do not. World War I and II are different in that the latter was a "moral war" with the "good" and the "bad" demarcated while the former was not. When Spielberg sets a film in World War II, it is a war film (*Saving Private Ryan*, 1998) but World War I produces an anti-war film (*War Horse*, 2011). But occasionally, the two wars can also yield the opposite results: *Catch 22* set in World War II being anti-war while *1917* (2019) is a war film. Some wars, when revisited can also be treated differently as with the Vietnam War. *The Green Berets* (1968) was a war film while Coppola's *Apocalypse Now* (1979), Oliver Stone's *Platoon* (1986), and Kubrick's *Full Metal Jacket* (1987) are anti-war.

Paths of Glory (1957)

The film begins in 1916, soldiers being marshaled in a chateau converted into what looks like the military headquarters; the legend on the screen is about the war in the trenches and the stalemate, neither side making progress of more than a few yards at a time. General Georges Broulard (Adolphe Menjou), of the French General Staff, asks his subordinate, the ambitious General Mireau (George Macready), to send his division on a suicide mission to take a well-defended German position called the "Anthill" the day after. General Mireau refuses on the grounds that it is impossible and the lives of his soldiers are worth more to him than any personal advancement. But when Broulard mentions a promotion, his tone shifts to praise of his men, nothing being impossible for them and turning this to accepting the task. Thc beginning of the film, since it implies duplicity through quickly shifting attitudes when the lives of so many men are involved, is satirical.

When General Mireau passes down the order, the officer designated with the duty to carry the attack on the "Anthill" is Col. Dax (Kirk Douglas), a former criminal defense attorney who knows that the task is suicidal but still cannot but accept it as the mission. As a first step, three soldiers led by the drunken Lieutenant Roget are sent on a reconnaissance mission and, due to his cowardice, one of the others Corporal Legeune, is killed. Corporal Paris, the other survivor, was once Roget's classmate and there is hostility between the two. Since an officer's word is more likely to be accepted, Roget falsifies the report without the Corporal resisting.

The attack on the "Anthill," when it takes place, is a failure and a company is unable even to leave the trenches. General Mireau, incensed at this, orders the artillery to fire at their own positions, which does not happen because the artillery commander demands a written order that is not forthcoming. As a punishment to the entire regiment for cowardice, General Mireau, in consultation with General Broulard, decides that one soldier from each of the three companies will be court martialed for cowardice as an example to the others. Three soldiers are chosen; Lieutenant Roget picks Corporal Paris and another is picked at random. Colonel Dax defends the men but, after a one-sided trial, the three are sentenced to death by firing squad. The night before the scheduled execution Dax confronts Broulard at a ball, with sworn statements by witnesses attesting to Mireau's order to shell his own trenches, in an attempt to coerce the General Staff into sparing the three men. Broulard retains the statements for further use but perfunctorily dismisses Dax, returning to the ball on the floor below. The three men are executed, one of them not even conscious, since he fractured his skull the previous day in a tussle.

Following the executions, Broulard breakfasts with Mireau and he has invited Dax to attend. He lets out that Mireau will be investigated for the order to fire on his own men. Mireau storms out, protesting that he has been made a scapegoat, threatening Dax of insubordination. Broulard then offers Mireau's command to Dax, admiringly assuming that Dax's attempts to stop the executions were a ploy to gain Mireau's job. Discovering that Dax was sincere, Broulard rebukes him for his idealism, while the disgusted Dax calls Broulard a "degenerate, sadistic old man," something that does not affect Broulard, amused that Dax actually wanted the three men saved. The film concludes shortly afterward with the men reveling at an inn listening to a German girl sing, "The Faithful Hussar." Their jeers turn to empathy as they begin humming with her and then join her in the singing. Colonel Dax, listening from a distance, receives news that the soldiers are due back at the front but he instructs the orderly to let them sing for a few minutes more.

Paths of Glory is a powerful film and its influence stems from the fact that it moves up and down the military hierarchy, portraying motives and methods from the top generals down to corporals and sergeants.[5] Apart from the hypocrisy of the generals, Lieutenant Roget, who must fortify himself with liquor before a mission panics while on reconnaissance, kills Corporal Lejeune and then falsifies the written report. Paris, who appeared principled in challenging Roget, in the end acquiesces in the presence of a superior officer but is still chosen for execution. The film is like the *Dr Strangelove* in its complete transparency, the certainty of the motives, most of them ignoble. Only Colonel Dax's sincerity makes it different; but still, Group Captain Mandrake in *Dr Strangelove* was sincere and truly did his best to avert a nuclear catastrophe. Kubrick's gaze is hence satirical and only the conclusion distracts us from that. But the response of the soldiers to the German girl is an add-on to the plot. Most of the film takes Colonel Dax's perspective and deals with his acts but the episode is only witnessed by him; he could have witnessed something else equally to serve as a conclusion.

This brings us to the central figure of Colonel Dax which is differently drawn from the rest of the characters in that where the others are all symptoms of human weakness, even those who are victims of injustice, Colonel Dax is representative of how human beings "should be." Apart from his moral stand vis-à-vis the soldiers he is defending, he leads the advance against the enemy under heavy fire, not displaying the fear that the others are showing and seemingly immune to the barrage from the other side. But the question here is that if Colonel Dax acts thus and still does not suffer, what justification is there for the other men to display signs of "human weakness?" One does not know the terms under which Kubrick made the film (since it was early in his career)[6] but it is a satirical vision subverted by the presence of a leading star in an implausibly heroic role, and the posters of the film similarly privilege Kirk Douglas. *Paths of Glory* concludes with Colonel Dax being protective toward the men who are due back on the front, further evidence of the influence of a star attending to his liberal image.[7] What is ultimately offered is the picture of one heroic man standing amidst the devastation that is humanity. *Paths of Glory* is apparently a mixed representation testifying to both its director's naturally satirical sensibility and its male star's pursuing of his liberal image.

Full Metal Jacket (1987)

The film is essentially an account of a group of recruits undergoing training to be Marines in Parris Island, South Carolina, following one of them J.T. Davis (Mathew Modine) into the Vietnam War. It begins with the recruits having their heads shaved, ridding them of individuality, as it were, and then being put through the mill. The drill instructor Sergeant Hartmann (Ronald Lee Ermey) is harsh and begins the training by telling the recruits that they are all "scum" and that the training will make them "killers." Each of them is given a nickname as a further effort at reducing them in stature. Three recruits are important here, the narrator Davis is given the name "Joker" and the overweight Leonard Lawrence (Vincent

D'Onofrio) is called "Pvt Pyle" after a dim-witted mechanic in a TV show and John Rucker becomes "Cowboy."

The training is grueling and the instructor does everything in his power to lower the self-esteem of the recruits through invective and punishment. Pyle finds himself unable to keep up with it and, to make matters worse, a jelly doughnut is discovered in his locker; Sergeant Hartmann decides to punish the whole group for Pyle's indiscretion, making them resent him. The turning point is reached at a "blanket party"—the entire platoon securing Pyle with a blanket and assaulting him with knotted towels containing bars of soap. Joker also joins in although reluctantly and this is hard because he had been assigned to look after Pyle. A change comes over Pyle after this and he starts talking to his gun, although he excels at marksmanship. They graduate and each of them is given an assignment, Joker as a military journalist and the others to the infantry. On their final night on Parris Island, Joker discovers Pyle in the latrine loading his rifle and executing drill commands, and loudly reciting the "Rifleman's Creed." This awakens the entire platoon and the drill instructor Hartmann who confronts Pyle and orders him to surrender the rifle. Pyle shoots Hartmann dead and then shoots himself in the mouth.

In the two films examined hitherto, Kubrick does not show himself to be interested in the thinking of his characters, in even the smallest with their feelings and this is also true of *Full Metal Jacket*, where there is no intimate engagement with anyone. The relationship between the recruits and the instructor resembles the one in *An Officer and a Gentleman* (1982) but Sergeant Hartmann makes the drill instructor in that film (Louis Gossett Jr) look benign. Still, there is, unlike in that film, no sense of what the protagonists are feeling, no private conversation, for instance, where Pyle might have revealed something of himself to Joker that might have humanized them both. The opening haircut sequence is a brutal look at what military training does to individuals but Kubrick does not show much actual interest in his recruits' individualities.

This first part of *Full Metal Jacket* sees Kubrick's irony in full flow because of Kubrick's distance from his characters but how does one identify it as irony?

> Most (theorists) agree that the degree of ironic effect in a text is inversely proportionate to the number of overt signals needed to achieve that effect [...]. But signals must perforce exist within the text in order to allow the decoder to infer the evaluative intent of the encoder. And irony is usually at someone's or something's expense.[8]

Irony cannot be continuously present in the entire text and can only function when there are parts from which it can be distinguished. In *Paths of Glory*, for instance, the segments featuring Kirk Douglas as Dax expressing his indignation are not ironic and nor is the conclusion with the soldiers singing.[9] The "sincere" parts exist in opposition to the ironic segments as when the generals are talking, their conversation is completely at loggerheads with the segments where the ground-level effects of their duplicitous talk are shown. In the same way, the irony in the first part of *Full Metal Jacket* is made evident largely by the second part, although not completely so. The contrast between the first and second is the contrast between creating "killing machines" in the laboratory and having them fight in real circumstances, and that is the primary irony. But there is the secondary irony in the first part alone when we see the weak humans being "processed" into mechanical monsters that they are unfit to become. Pvt. Pyle is only the extreme representative of the all too human young recruits under training.

In the second part of *Full Metal Jacket*, Sergeant Joker is in Vietnam as a war correspondent located at Da Nang on the eve of the Tet Offensive. The day after the unsuccessful attempt by the enemy to overrun the base Joker is sent to Phu Bai, with Pvt Rafterman to accompany him. There they meet the Lusthog Squad of which his old friend Cowboy is a member. Joker accompanies the squad during the Battle of Huế, where the platoon commander is killed by the enemy. After the Marines declare the area secure, a team of American news journalists and reporters enters Huế to interview various Marines about their experiences. While patrolling Huế, Crazy Earl, the squad leader, is killed by a booby-trapped doll, leaving Cowboy in command. The squad is targeted by a sniper from an unknown location who severely wounds two of the men, then using them as bait to lure the others in. Cowboy is also killed but the sniper is located and found to be a teenage girl. Squad leader Animal Mother proposes to leave the wounded girl to rats but Joker kills her on her agonized plea. It concludes with the squad returning to their Perfume River base singing the Mickey Mouse March, Joker in a voiceover saying he is glad to be alive.

One of the concerns attributed to Kubrick's films is the sense of rationally devised systems breaking down.[10] This is true of *Dr Strangelove* and of *Full Metal Jacket*, as well. In *Full Metal Jacket*, the system by which recruits are trained to be perfect killing machines collapses when Pyle has a murderous breakdown. But the breakdown of a system created and inhabited by humans needs to affirm the individuality of these humans and distinguish between their failure as human beings and the failure of the system devised. By not being sufficiently interested in people as people within the system—or in their fates—Kubrick does not explore the issue of the "systemic failure" as a notion as he might have. Here is Kubrick on the subject of "Man":

> Man isn't a noble savage, he's an ignoble savage. He is irrational, brutal, weak, silly, unable to be objective about anything where his own interests are involved and any attempt to create social institutions based on a false view of the nature of man is probably doomed to failure.[11]

But a question that naturally follows is that if it is humans who are so flawed, why should the collapse of the *systems they devise* be warranting of condemnation? Human achievements deserve criticism only if humanity itself (even if not "noble") is in pursuit of excellence, for only then can the achievements themselves be worthy or unworthy and judged correctly on that basis.

With the creation of perfect killing machines in the first part of *Full Metal Jacket*, the second part can be described as this expected perfection losing its way in the chaos of Vietnam. The contrasting milieus: Parris Island with its impeccable order and Vietnam with its randomness is an analogy for what Kubrick is trying to show: perfect equipment failing in a real situation. But just as Pyle's collapse would be more telling if he had been adequately individualized, the American military's collapse would have been more interesting with the soldiers had been humanized, shown as personally disoriented in such a chaotic milieu. Joker might have, if he had been different, been a moral counterpoint to the system, but he behaves like the others with Pyle. As in *Dr Strangelove*, Kubrick is taking his position coldly above humanity through the approach.

But there is little evidence that Kubrick even cared about the "human truth" in Vietnam, so preoccupied is he with "Man as ignoble." Elsewhere Kubrick says that any system devised by humans should take their darker sides into account.[12] If the "darker side" is self-interest and greed, one could ask if the market economy had not already done that but with human experience still being valued in the arts. A system devised by humans is fascinating only because it is inhabited by flawed humans still striving for perfection, and in Kubrick's above films, there is scant evidence of that. The "goodness of Man" may be empty rhetoric but an artist might need to still take his or her place among human beings; Kubrick's perspective, it would again seem, is too lofty.

The other films

I have tried to demonstrate that all three of Kubrick's films examined in the earlier sections are deeply ironic, with war and violence as key motifs. Kubrick's most highly regarded film is *2001: A Space Odyssey* (1968), mainly for its visionary qualities, and there could be resistance to it being termed irony or satire in intent. This section is devoted to three of his other films including *2001: A Space Odyssey*, none of which will perhaps pass by consensus for satire. But my argument is that Kubrick has a deep, generally ironic attitude toward humankind's ways that led to a satirical turn in his films, and that such a perspective would have been hard for the filmmaker to abandon regardless of his chosen subject matter. I will begin this section with an examination of *2001: A Space Odyssey*.

2001: A Space Odyssey (1968)

The film begins in a prehistoric veldt among a group of hominids, apparently humankind's ancestors. There is another tribe shown and there is a rivalry with them, but the two keep their distance. One day the first tribe senses something not seen before and we are shown what it is. A rectangular black monolith appears before them and, since this is "the dawn of man," it must evidently be extra-terrestrial. The monolith has a strange effect upon them and, very soon after this, one of the hominids begins to use a bone as a weapon, killing herbivores before his tribe attacks their rivals with it and notches up its first hominid kill. In exultation, one of the hominids throws up the bone into the air and we see it turning in mid-flight to a similarly shaped spaceship.

Kubrick is making a statement here about humankind's progress from the moment when it began to use tools up till today, but it is significant that their use connected to technology today is its first use in war. This connects up with *Dr Strangelove* in that technological progress is tied up to destruction and war. By locating the scientific/technological impulse in the urge for war, Kubrick is dealing with an "original sin" in the hominid state, that technology was born inherently corrupted by the urge to kill one's own, i.e. the war impulse that develops into *Dr Strangelove.* It is also significant that the hominids acquire technology through the intervention by extra-terrestrials. If we consider the notion of extra-terrestrials, can their innate impulses be different from that of humans since they must have developed in comparable conditions? Given this line of thought, it may not be a matter of chance that alien intervention first produced war in *2001: A Space Odyssey*.

Leading up to the year 2001, Dr Heywood Floyd, Chairman of the United States National Council of Astronautics, travels to Clavius Base, a United States outpost on the Moon. During a stopover at Space Station 5, he meets Russian scientists who are concerned that Clavius seems to be incommunicado but refuses to discuss rumors of an epidemic at the base. Continuing his journey, Floyd addresses a meeting of personnel to whom he stresses the need for secrecy with respect to their newest discovery. Floyd's mission is to investigate a recently found artifact buried four million years ago near the crater Tycho. Floyd and others proceed to the artifact, a black monolith identical in shape to the one encountered by the hominids. As they examine the monolith, it is struck by sunlight, upon which it emits a high-powered radio signal. Eighteen months later, the United States spacecraft *Discovery One* is bound for Jupiter. On board are mission pilots and scientists Dr David Bowman and Dr Frank Poole, along with three other scientists in suspended animation. Most of *Discovery*'s operations are controlled by a HAL 9000 computer with a human personality and many human impulses.

Before proceeding further with the account of the film, it would be pertinent to look at the way Kubrick presents the technological future. At first glance, he is scrupulously attentive to likelihoods about space travel—like grip shoes to make sure people do not suffer the unpleasant results of weightlessness, food coming in sealed container from which it has to be sucked out and zero gravity toilets. But the conspicuous factor is the banality of the representation including the non-communicative nature of Dr Floyd's brief talk, attended by the press and caught on camera. Since we are aware that there is sensational news hidden behind all this, Kubrick may be simply creating the right atmosphere for the forthcoming revelation, but there could be something else also. The conversation is empty of significant content, not only Floyd's speech but also his talk with his little daughter. The choice of actors is significant and there is an attempt to make them conform to a few physical types. Dr Heywood Floyd (William Sylvester) and the man who introduces him before his talk Dr Ralph Halvorsen (Robert Beatty) are so, as are the astronauts Bowman (Keir Dullea) and Poole (Gary Lockwood). They are all clean-shaven but as difficult to distinguish from each other as the hairy hominids.

The penultimate part of the film is on the spaceship heading toward Jupiter with Bowman and Poole gradually finding HAL to be malfunctioning, although "insanity" would be a better description. HAL's conversation is actually more human than that of the humans in the film since it expresses wonder with phrases like "but the odd thing was" implying that HAL has something corresponding to an imagination—that the humans are not shown to possess. HAL becomes suspicious when Bowman and Poole discuss it inside a pod and lip reads their talk to conclude that his own existence is threatened. HAL is shown through an unblinking red light that corresponds to a madman's stare. By some subterfuge, therefore, the paranoid HAL manages to shut both Bowman and Poole out of the spaceship and Poole is killed as are the hibernating astronauts, "terminated" in their unconscious states. Bowman however manages to get back inside and disconnects HAL, who "dies" (instead of being "shut down") singing "Daisy, daisy, give me your answer."

The contrast between the men and HAL is striking and it is also difficult to assign a rational reason for why HAL turns paranoid. My own interpretation is that in the future, technological humankind will, through AI, transfer its own basest impulses to its "thinking machines." The first hominid who kills (named "Moonwatcher") becomes "human" through the urge to kill[13] and HAL 9000 feels it as well. But given this aspect of *2001: A Space Odyssey* till this point, what are we to make of the last visionary segment, Bowman going "beyond the infinite" to become the Star Child? For Steven Spielberg, the Star Child at the end of *2001: A Space Odyssey* is "the greatest moment of optimism and hope for mankind that has ever been offered by a modern filmmaker."[14]

In the last part of the film, as *Discovery One* approaches, Jupiter Bowman finds a third monolith orbiting the planet. He leaves *Discovery One* in an EVA pod to investigate but is pulled into a vortex of colored light. Bowman is carried across vast distances of space while viewing bizarre cosmological phenomena and strange landscapes of unusual colors and the sequence continues for a full ten minutes before Bowman's pod is deposited in a neoclassical bedroom where he sees, and then becomes, older versions of himself, finally as an old man lying on a bed. A monolith appears at the foot of the bed, and as Bowman reaches for it, he is transformed into a fetus enclosed in a transparent orb of light, which floats in space beside the Earth. Given the structure of the film, we could argue that the narrative stops with the announcement of the section titled "Jupiter and beyond the infinite" since what follows is pure spectacle and special effects. Since a key piece of music in the film is Richard Strauss's "Thus spake Zarathustra," the reading of the creation of the "Star Child" is that it represents Frederic Nietzsche's Übermensch (Overman),[15] which finds a place in the philosopher's 1883 book *Thus Spake Zarathustra.*

This brings us to ask what Kubrick could be meaning, given his pessimism, but there is an insight to be found here, which is that:

> Kubrick's fatalistic view of humankind is always less political than anthropological— even in *Paths of Glory*, *Dr. Strangelove*, and *Full Metal Jacket*, which, while dealing with atrocities that might be deemed politically preventable, imply that all of this has happened many times before, and will take place again, and yet again.[16]

Once we look at his films this way, the Star Child in *2001: A Space Odyssey* is wish-fulfillment it would seem, and only for a brief moment at that. Having no reason to expect humankind to correct itself on its own, he is positing "outside intervention" although there is no reason to suppose that extra-terrestrials will not also be handicapped with mankind's flaws, or other ones equally debilitating; their intervention, after all, created a hominid war. Moreover, if this is "hope" for Kubrick, we must also take into account that, considered chronologically, his films hardly got more optimistic and *Full Metal Jacket* came much later. In this context, the conclusion of *2001: A Space Odyssey* emerges as mimicry of religious belief—although one might also debate if this mimicry is actually a parody and the alien intervention laden with irony. But regardless of how we interpret *2001: A Space Odyssey*, the idea that humankind can be saved only through external intervention is hardly convincing as "hope." *2001: A Space Odyssey* leaves much room to be misunderstood but that is as true of the next two films. But one thing we may be convinced of is that Kubrick's vision of humankind's future is not ecstatic but tinged with irony.

The Shining (1980)

Kubrick's film of Stephen King's novel *The Shining* belongs to the horror genre but it seems to do much more and has been interpreted in various ways including as a reflection on the Holocaust. He had worked on a screenplay entitled "Aryan Papers" on the subject of the Holocaust but was unable to film it:

> Part of Kubrick's reluctance to deal with the Holocaust directly in his work had to do with the aesthetic problem of bringing art to bear on such a subject. The appalling horror of the Holocaust made it a particularly difficult subject for a satirist like Kubrick. The Holocaust was also a threatening topic not only because Kubrick was Jewish but because of the particular nature of some of his social, philosophical and historical concerns. Unlike the rational systems whose breakdown Kubrick habitually contemplates in his films, the Nazis' Final Solution was not a rational system gone wrong, it was a rationalized system gone horribly right. Moreover, the Holocaust at its black core is a horrible mystery of irrational evil that has in new and extremely powerful ways thrown the nature and even the existence of civilization and God into question.[17]

Critics have generally based the association of *The Shining* with the Holocaust on flimsy evidence but the general approach is to treat the site of the action—The Overlook Hotel in the Rockies—as a site of great past violence that the boy Danny alone is able to perceive because of the quality he possesses the "shining." To demonstrate the arbitrariness of such a reading (largely dependent on Kubrick being Jewish), another sees *The Shining* as a fairy tale[18] using Bruno Bettelheim's *The Uses of Enchantment* based on the information that Kubrick and script collaborator Diane Johnson read it while preparing for *The Shining*. My reading of the film will overlook these interpretations since the film itself offers scant evidence of such relationships and they depend on extra-textual information. Kubrick was an acknowledged great of cinema and he deserves to have his *texts examined on their own* for what they might reveal. Even if he seemingly follows the text he is adapting, he transforms it with his mise en scène.

Set in contemporary America, King's novel is essentially a haunted house thriller with obvious ties to eighteenth- and nineteenth-century Gothic traditions. In the film, Jack Torrance (Jack Nicholson), a former schoolteacher eager to become a writer, takes a job as the winter caretaker of the Overlook Hotel, an imposing and somewhat old (constructed in 1907) Colorado mountain resort brought up to date. Jack's wife, Wendy (Shelley Duvall), is fearful of her husband, a recovering alcoholic, but eager for the change. Their young son, Danny (Danny Lloyd), was once physically abused by his father. Danny possesses psychic powers termed "shining" and has created an imaginary playmate named Tony, who appears to furnish him with knowledge about the past, present, and future. Danny "shines" back to a gruesome moment in the hotel's history that frightens him: the previous caretaker, a man named Delbert Grady, murdered his wife and two daughters with an axe before shooting himself in the mouth. At the hotel, Danny meets briefly with the departing cook, Dick Halloran (Scatman Crothers), who shares his gift of shining and warns Danny about the presence of malevolent spirits and ghostly "traces" in the hotel.

Kubrick sticks broadly to the novel although he has apparently his own agenda—not pertaining to inducing belief in the supernatural—which means that there are elements impossible to interpret plausibly in the context of a horror novel. There are several ghostly presences in the film that begin with the bartender Lloyd, whom Torrance seems to know from before, Delbert Grady and his two daughters who appear separately, a naked woman in the bathtub of Room 237 who reportedly tried to strangle Danny, and who metamorphoses into a hag on Jack Torrance's touch. What Room 237 stands for is also not made clear. We are not allowed to regard the supernatural elements as visions, dreams, or illusions since Jack Torrance, when murderous and locked into the store room by Wendy, has Delbert Grady unlocking the door and setting him free. Ghosts do not normally perform such material functions even in genre stories. Then there is an indication that Jack Torrance has a long association with the hotel and we see his face in an old photograph. The hotel was constructed on the site of an old Indian burial ground and there are Indian artifacts on display. Lastly, Danny has a vision of torrents of blood flowing out of the elevator that is not explained. Hence, rather than interpret all these elements, I will simply examine the focal characters and the striking mise en scène representing the Overlook Hotel as Kubrick imagined it.

Beginning with Jack Torrance, he was a schoolteacher but has been seeing hard times since there is mention of his having worked in a car wash. Employment at the hotel is a golden opportunity for him since he and his family has never seen such luxury. Torrance harbors writing ambitions and at the climax of the film, Wendy chances upon what he has been typing: pages and pages of "All work and no play makes Jack a dull boy" on reams of white paper; there is perhaps no comparably horrific picture of writerly frustration in all of cinema. Wendy is, in contrast, a picture of utter banality, coming out only with vapid pronouncements. Shelley Duvall, who plays her, was in a key part of the same kind in Robert Altman's *Three Women* (1977) although her portrayal there was much more affectionate. Here Kubrick's portrayal of Wendy is nothing short of savage,[19] and Jack Torrance's conduct toward his wife emphasizes this at every stage, his words steeped in sarcasm that only grows worse. Wendy's portrayal depends on Shelley Duvall's physiognomy, which could make the viewer uncomfortable, and the charge of misogyny directed by Stephen King at Kubrick was not unfounded. Still, Kubrick is trying to reveal Jack's writer's block through the only muse he is allowed, which is Wendy. That this muse is played by Shelley Duvall is itself ironic.

Coming to the mise en scène there are two aspects of pertinence of which the first is the *horror of plenty* created, the quantity and items of frozen food in the store room where "one could eat for two years and still not have the same menu twice." As already elaborated upon Kubrick was taken up with the failure of perfectly devised human systems and the Overlook Hotel could be one such system.[20] Like Parris Island in *Full Metal Jacket*, it is intended to secure human beings for all contingencies. But it fails with ordinary people like Delbert Grady and Jack Torrance. That both of them respond in identical way implies that it is the system that is responsible. Seen in this context, the ghostly elements are simply like manifestations of the imperfect system just as, in *Full Metal Jacket*, the conduct of their Vietnamese allies—who are not aligned with the military purpose of the United States—manifests the failed military system. One way of characterizing the essential problems of the Overlook Hotel as an imperfect system is that while making itself perfect it neglected to take into account the past of the space. The hotel in its up-to-date-ness apparently erased this past. If the system in *Full Metal Jacket* cannot contain the weaknesses of the recruits, the Overlook Hotel cannot contain the intruding traces of the past that it has tried to erase or only include as decorations. The "shining" may then be the supernatural faculty that registers these imperfections as ghostly traces.

The second aspect is the comparison of the hotel with the maze outside, Danny tricycling wildly in the carpeted corridors and especially the use of color in the interiors: with the Gold Room where Jack gets his drinks from Lloyd and green and red dominating some other spaces. There are also the design motifs on the carpets in the corridors where a conspicuous motif is of an incomplete hexagon leaving a mouth open, suggesting an enclosure where people seeking shelter might be trapped.[21] The association of color with seclusion reminds us of Poe's story "The Masque of the Red Death" about a prince who, with his companions, shuts himself off from a plague in within the opulence of a castle, in seven rooms of different colors.

Lastly, an issue is why Danny and Wendy survive and it is Jack Torrance who is killed, and a clue can perhaps be found in Grady telling Jack that Wendy is more resourceful than they had imagined. Wendy and Danny can negotiate their way through the maze but Jack cannot; still, despite the two being saved and Jack dead, the ending does not emerge as a "happy" one. If we consider the terrifying sequence in which Jack follows Wendy up the stairs, she armed with the baseball bat we find that Jack is being rational in what he is saying in virtually every sentence he utters, although his intense desire to kill Wendy for her incomprehension of his state is not. An illustration is his denial of time to Wendy to think things over: "You've had your whole life to think things over; what good would a few more minutes do you now?" It would seem from these factors that Jack is the victim in a system to which Wendy is perfectly attuned.

If we consider Jack Torrance as a failed writer and alcoholic, the image virtually corresponds to the modernist cliché of the self-destructive writer still devoted to his vocation. But the milieu in which he is trying to be productive is, regrettably, postmodern America, marked, as theorists describe it, by the very triviality of life in late capitalism, with the ethic of conspicuous consumption replacing that of production, brought alive by the excesses of the Overlook Hotel:

> If the ghost story [...] usually foregrounds the destructive designs of malevolent spirits on the living, *The Shining* emphasizes the vapidity and psychological fragility of the present. Trapped by marriage and the responsibility of supporting his family, Jack Torrance retreats to a grand emptiness that he hopes will afford him the semi-solitude he needs to realize his ambition of becoming a writer [...] Jack Torrance thus becomes the perfect image of the failing artist in postmodernity: trapped between times, unable to find within or without anything worth saying, reduced to the exercise of primitive, pointless resentment at the bleak prospects of the human condition.[22]

We could, in contrast to Jack Torrance, regard Wendy as a subject of postmodernity, which explains her survival, though Danny's survival perhaps owes to the horror of killing children in films.[23] We must also consider the significant aspect of Wendy being the only one who is virtually immune to ghosts although, according to Jack Torrance, she is a "horror film and ghost story addict." The ghosts, we recall, manifest the failed system and Wendy, in her vapidity, can detect no failure there.

There are many inconsistencies in *The Shining* and has confused people but it also seems to contain a more complex social statement from Kubrick than any of his other films examined hitherto. A key reason may be that it is dealing specifically with a social symptom of the West and not the broad category of "humankind"; unlike *Dr Strangelove* or *2001*, we cannot ask where Kubrick is viewing everything from. An artist cannot legitimately take his or her position outside humankind since art is social communication of a sort but he or she can nonetheless be an outsider to a particular culture. *The Shining* and *Eyes Wide Shut*, which are Kubrick's most complex and puzzling works, can both be read as commentaries on a specific milieu. Both have invited disagreement among critics and deserve the greatest attention in his oeuvre.

Eyes Wide Shut (1999)

Eyes Wide Shut, which is set in New York, begins with Dr Bill Harford (Tom Cruise) and his wife Alice (Nicole Kidman) preparing to leave for a party thrown by a wealthy patient Victor Ziegler. They leave their darling little daughter Helena in the care of a babysitter on their way out. Dr Harford is evidently a successful professional but their host is much richer and Alice wonders why they keep getting invited each year to Ziegler's party where they know no one else. Harford however discovers that he knows the pianist Nick Nightingale who dropped out of medical school when they were studying together, to pursue music. In the course of the evening, an aristocratic Hungarian named Sandor Szavost dances with Alice, attempts to seduce her when she is under the influence of champagne while her husband is similarly accosted by two young women—though he is diverted by an attendant to a private room in which their host is having a crisis. A young woman with whom Ziegler was having sex has overdosed on a combination of heroin and coke. Bill Harford revives her and gets Ziegler's gratitude.

Back in their apartment an evening later, the couple discuss their respective sexual experiences over marijuana. Alice wonders if her husband would have gone along with the two young women if he had not been interrupted, and his remonstrations that he has only eyes for Alice—because he loves her—gets him deeper into trouble. We are familiar with Tom Cruise's screen persona and Bill declines to admit even imagining himself with other women because of his deep love for his wife. Alice is acidic and the scene concludes when she admits to having been so infatuated with a naval officer the previous summer that if only the man had wanted her, she might have given up everything. Unlike Bill's love for Alice, hers for him does not preclude sexual fantasies about other men. It is at this point that Bill receives a call telling him of the death of a patient and he is required to proceed to the man's house post-haste.

When Bill is led to his dead patient, he finds the man's daughter Marion alone with her father's corpse. Her boyfriend Carl, who she is due to marry the next year, is not present, and Marion makes use of the opportunity to kiss Bill fervently and declare her love for him ("I love you; I love you; I love you!"). Like other, similar sequences in the film Kubrick stages it acerbically. What is to be revealed is held back while the person voicing it hesitates, seems embarrassed at the inappropriateness of the utterance before abruptly plunging ahead. In many of these sequences, a woman makes sexual overtures to Bill, who is plainly ill-equipped to deal with them because of his moral baggage—that could be associated with Tom Cruise's screen persona. If these sequences are wicked in their comic appeal, it is not only because Bill's prudishness is being mocked; the star's lily-white screen persona is itself being made to look foolish. Tom Cruise plays each scene in his customary way which, given the altered context, is incongruous to say the least. But Kubrick deliberately enlists this discomfort to illuminate his protagonist's naiveté, and there is little evidence that the star understood this aspect of his role. Evidently, there is irony in this portrayal of a lily-white character being made to look foolish *played by Tom Cruise.*

Bill Harford leaves Marion when Carl comes in and in the next sequence finds himself accosted by a hooker and accompanying her to her apartment. The two discuss the price and the corresponding services offered and $150 is agreed upon circuitously with the exact services left to be decided by Domino the hooker. But Harford gets a call from Alice at this moment and *Domino* decides that *he* wants to leave. The doctor nonetheless gives Domino the promised $150 much against the girl's protests. A walk along the pavement outside and the crossing of a few intersections leads Bill to the cafe in which Nick Nightingale is playing and this turns out to be a most fateful event. Nick Nightingale reveals that he is due to perform elsewhere that night and there is a secret gathering of masked and costumed people and Nick Nightingale is required to play there blindfolded. These meetings have happened many times before but it is always in a different place. One time the blindfold slipped and Nick saw things he had never seen before—especially the array of women on display. All this is revealed by Nick when a call comes to give him a password, which is "Fidelio" this time. Bill knows the password now and he is insistent that he wants to come along. Nick finally consents on the condition that he gets there on his own, but Bill needs to get a costume and a mask.

It is now very late but Bill Harford drives to a place that rents out costumes. The man he knew there has moved out but the shop is now owned by one Milich, who agrees to rent him a costume for the usual fee plus the $200 incentive offered by Bill. But just before the transaction is concluded, Milich discovers his underage daughter half-naked with two middle-aged Japanese men in drag, hiding behind the furniture. Milich is now caught between concluding the transaction with Bill and notifying the police about the men but, when he is dealing with both issues together, his juvenile daughter nestles close to Bill, making her intentions apparent. Bill, of course, remains unmoved by this new turn and he gathers his costume and leaves—for the highpoint of the night. It is nearly 2 a.m. and he is armed with the password "Fidelio."

The gathering that Bill Harford wishes to infiltrate is being held in a mansion some distance away and he takes a cab to get there. The cab driver agrees to wait for an extra inducement of $100 and Bill is finally conducted into the exclusive masked gathering, suitably attired and disguised. What is underway in an enormous, dimly-lit space is an orgy of some sort. When Bill enters the master of ceremonies is chanting something in a strange language while Nick Nightingale plays bars of music blindfolded on an electric organ. The master of ceremonies is encircled by cloaked and masked women who soon shed their cloaks, revealing themselves to be naked except for G-strings. Bill tries to blend into the gathering but, in no time at all, finds he has been spotted as an intruder. A naked woman in a mask lets him know this first but there have also been other gazes fixed on him. Within a few minutes, he is summoned by the master of ceremonies and unceremoniously sent out, the only consolation being that he is not stripped naked because of the intervention of a masked woman, who takes responsibility for his future good behavior.

Back home Bill Harford finds his wife fast asleep but restless and giggling. When she is woken up roughly, Alice tells him of the "nightmare" she has been having but there is a hint that she is not telling him the truth about the dream but concocting an elaborate one he might be more comfortable with. Bill returns the costume to Milich but he is now cheerfully transacting with the Japanese over his daughter. The rest of the film has to do with the threats made to Bill when he tries to investigate, his visiting Nick Nightingale at his hotel but finding that he has disappeared in the company of two men, Bill visiting the hooker Domino but being informed—with the now familiar circuitousness—by an equally solicitous roommate of Domino having tested positive for HIV. Bill also learns from a newspaper that a former beauty queen was found dead from a drug overdose. He suspects that this is the girl from the previous night and this is later confirmed. Just as he is beginning to have suspicions of a monstrous conspiracy closing in on him, Victor Ziegler calls him over and explains some things to him.

"What were you trying to do last night?" Ziegler asks him. Ziegler was present at the gathering and saw what happened. He also tells Bill that he was completely out of his depth and he would understand this if he knew the names of some of the others present. "You would lose sleep over it," Ziegler lets him know. Also, Bill's dramatic expulsion from the party, with the girl coming to his rescue, was a charade arranged to put some fear into him. Nick Nightingale did not "disappear" but was put on a plane back to Seattle. As for the girl who overdosed, she was the same hooker that Bill saved earlier. Her death was unfortunate but she was an addict and Bill knew very well that it was inevitable. Bill is left speechless by Ziegler and the film ends with his returning to Alice and Helena and telling his wife everything. Husband and wife wonder if it was all a dream but there is still a promise of conjugality for the near future.

Stanley Kubrick's film was inspired by Arthur Schnitzler's 1926 novel *Dream Story* (*Traumnovelle*), set in Vienna in the 1890s, which explores the erotic fantasies of a socially successful Viennese doctor, Fridolin, and his wife, Albertine. Kubrick sets the film in contemporary New York but it has been pronounced as remarkably faithful to Schnitzler's book.[24] Kubrick's own remarks about the film do not contradict that but neither did he deny that *The Shining* was in any way a different object from Stephen King's novel. One suspects that he did not entirely share his intentions even with his collaborators, notably the actors. Casting Shelley Duvall in *The Shining* may have been a misogynistic (though brilliant) decision based on her physiognomy but he perhaps kept the reason for casting her to himself. I would say that he cast Tom Cruise and Nicole Kidman (an off-screen husband and wife pair) in the same way here but, more than his physiognomy, for what Tom Cruise represents in American cinema.

Dream Story is an erotic novel and Schnitzler was a kindred spirit of Freud's and the first question one should ask is whether there was in it anything for Kubrick – who has never shown himself interested in such matters. A critic in the *New York Times* pointed out, Kubrick "never paid much attention to the psychology of characters, much less relationships between men and women," and in fact "spent his career ignoring (or avoiding) the inner lives of people, their private dreams and frustrations,"[25] leading us to further wonder both at his choice of the source and what he had done with it since most critics have been hesitant to interpret it and taken it as straightforward.[26] Slavoj Zizek, one of the bolder cultural commentators, has the following to say about it:

> Recall Stanley Kubrick's *Eyes Wide Shut* (1999): it is only Nicole Kidman's fantasy that truly is a fantasy, while Tom Cruise's fantasy is a reflexive fake, a desperate attempt to artificially recreate/reach the fantasy, a fantasising triggered by the traumatic encounter of the Other's fantasy, a desperate attempt to answer the enigma of the Other's fantasy: what was the fantasized scene/encounter that so deeply marked her? What Cruise does on his adventurous night is to go on a kind of window-shopping trip for fantasies: each situation in which he finds himself can be read as a realized fantasy—firstly the fantasy of being the object of the passionate love interest of his patient's daughter; then the fantasy of encountering a kind prostitute who doesn't even want money from him; then the encounter with the weird Serb (?) owner of the mask rental store who is also a pimp for his juvenile daughter; finally, the big orgy in the suburban villa. This accounts for the strangely subdued, statuesque, "impotent" even, character of the scene of the orgy in which his adventure finds its culmination.[27]

Eyes Wide Shut is nominally faithful to its literary source, a work with psychoanalytical implications, but there are nonetheless elements in it that make its impact nasty as comedy. The key element here is the casting of Tom Cruise in the central role, and making his image the covert target of ridicule. Apart from the segments I have already described, the circuitous way in which matters are explained to him, while he waits in open-mouthed vacuity, leaves the star at a distinct disadvantage. In fact, if a comparison were to be made, it is like Jack Torrance explaining matters in *The Shining* to his wife Wendy, made deliberately stupid. Kubrick also puts in more than one scene in which Bill Harford is taken to be gay and, considering that Tom Cruise has consistently played macho roles (*Mission Impossible*), this could be a deliberate slight on his masculinity. I am not certain that this is aesthetically legitimate but that the star is being mocked in the film has not gone unnoticed by other critics. Here is how the *Slant Magazine* review of the film begins: "Misunderstood as a psychosexual thriller, Stanley Kubrick's final film is actually more of an acidic comedy about how Tom Cruise fails to get laid."[28] Nicole Kidman, on the other hand, is hardly the butt of humor; a reason may be that the film is a take on American society and Kidman is an outsider (Australian). Moreover, her persona is more ambiguous (*To Die For*, 1995) than Cruise's.

But if making Tom Cruise's image the target of humor is the focus of the film, a question is how such a film can pass for "art"; is it not reasonable to expect "art" to have something of larger interest for us? The only way the project can make sense as art is to regard it in terms of what Tom Cruise represents. If one studies his oeuvre, one finds that the star has, generally speaking, been an emblem of middle-class aspiration. Many of his biggest hits have focused on young people of integrity aspiring to top their professions/vocations like *Top Gun* (1986), about an air-force pilot, *Cocktail* (1988), about a business student working part time as a bartender, *A Few Good Men* (1992) and *The Firm* (1993), which are both set among lawyers and *Jerry Maguire* (1996), about an honest sports manager.

In *Eyes Wide Shut*, Tom Cruise plays a successful doctor with many of the same characteristics. There are, in fact, several occasions on which he flashes his medical registration card at other people as if to reassure himself that his motives are strictly legitimate. To make his happiness complete, Bill's little daughter is as adorable as Hollywood ever made children. Bill also has plenty of money; his affluence and ease with money are things that the film is always emphasizing. Yet, his exclusion from another realm is more important and the first inkling we get of this is when Alice asks him why Ziegler invites them to his parties every year although Ziegler's circle is hardly their own. Bill's answer is particularly instructive, "That is what you get for making house calls," he says and it is not with any special irony. Later in the film Bill Harford tries to break into the orgy but he is immediately spotted, despite his disguise. There is no way in which even an affluent professional like Bill could belong in the elite and his immediate expulsion is a token of this fact.

Kubrick conveys something about the elite here which is also deeply mordant. The only legitimate participants at the orgy apart from the elite guests are apparently hookers. One could also argue that Bill's sexual squeamishness is being presented as an ideal of his class and it is this squeamishness that grates on Alice—when she is prepared to admit her own fantasies. Also, Bill is initially not receptive to Marion because of his "integrity" but, later, when he has been denied sexual gratification constantly, he tries to call her and her fiancé picks up the phone and he cuts it off. When Bill is finally cast out of the orgy, the master of ceremonies demands that he strip naked but he is saved by the woman's intervention. The observation is that a macho star being made to strip naked in this fashion is designed as a deliberate humiliation; since Tom Cruise's role as Bill Harford fits perfectly with his image, the object implicitly at the center of the humiliation is the star image as an American social ideal.

The film deals with several subjects which might be taboo in mainstream cinema—matters like child prostitution (done in an "unforgivably flippant" way). The use of a Serbian Milich as the child's genial father instead of an American is to make it more tolerable; the clients are also not American but Japanese. This act of "political incorrectness" may be placed alongside Bill being called a "faggot" roughly and without provocation by youth on a street and another where the male receptionist at Nightingale's hotel makes eyes at him. Another scene unthinkable to Hollywood is a wife making deep admissions to her husband (whom she loves) but the admissions being concocted. I already described the scene where she is giggling in her sleep; when she relates the dream to her husband, he is turned away from her and she is half smiling in a strange way, which is inconsistent with the admission of the "horrible dream."

Coming to what the "elite" means in *Eyes Wide Shut*, one could interpret it as the American plutocracy. It is a common understanding that there are powerful elites on behalf of which public decisions (including those pertaining to war) are taken; the way in which participants in the financial scandal of 2008 were let off and/or got key posts later only underscored this. But while the existence of this plutocracy is inferred, it still remains an elusive notion since how it operates can only be a matter of conjecture. The orgy in *Eyes Wide Shut* can hence be understood as representing a middle-class fantasy about the lives of this class—elite that one does not know much about. It is not a personal fantasy of Bill Harford (as Zizek would have it), but a fantasy that his entire class may be living in their secret moments. Kubrick uses a Romanian liturgy, even playing it backward, for the music in the scene to be especially weird and unlikely.

Kubrick's film can be interpreted fruitfully as a satirical comedy about an affluent middle-class professional trying unsuccessfully to enter the elite, which has none of the moral niceties that he himself—as a middle-class American—has. A statement attributed to Scott Fitzgerald is the following one: "The rich are not like you and me; they are different." Bill Harford is perhaps the kind of naive middle-class American who does not know that the rich are "different." "Eyes wide shut" characterizes his condition but as in Kubrick's *The Shining*, the satire in this film is hardly apparent. It thrives on the irony of Tom Cruise presented in the role of the icon of middle-class professional integrity that he has played, but with that role brutally mocked. Apart from the film's visual lushness which makes it too sumptuous for satire, there are also attempts at confusing the audience. Alice, who is so supremely bitchy/mischievous with Bill earlier in the film, is shown without make-up and deeply disturbed at the end; this is not consistent with the film being satirical. At the same time, the scene does not add up to anything else either. There are such inconsistencies in *2001: A Space Odyssey* and *The Shining* also suggesting that notwithstanding his visual impeccability Kubrick was not exact in his narrative ploys.

My understanding of the inconsistencies is that Kubrick in making the film for a wide American audience had no conviction that its meaning would be understood—because what was being satirized was not acknowledged by the public it was meant for. The film is a nasty attack on the morally solemn American middle classes, the solemnity of which has been induced by earnest myth-making (chiefly their moral sentimentality and "political correctness"), which the plutocracy that controls America could even be derisive of. It critiques America as a failed system, a failure Americans are not cognizant of, and does it by mocking its most vocal classes through their icon, although covertly. *Eyes Wide Shut* is perhaps riding on Kubrick's supreme craftsmanship and his stature as an auteur without the director even being particular that his audiences should respond to its nasty vision, chiefly a picture of themselves that they have never been shown. One cannot but feel ambivalent about a work of great artistic beauty if it shows no faith in being able to touch its audience as it might want to. But that seems a recurring characteristic of many of the films and filmmakers dealt with in this book.

NOTES

1. "Just as Hollywood silent cinema borrowed avant-garde devices but used them to narrative ends, so recent American filmmaking has appropriated art film devices [...]. The open endings of *2001* and *Five Easy Pieces* the psychological ambiguity of *The Conversation*, *Klute*, and *Three Women* testify to the assimilation of the conventions of the art film." David Bordwell, "The Art Cinema as a Mode of Film Practice," *Film Theory and Criticism: Introductory Readings*, ed. by Leo Braudy and Marshall Cohen, 5th edn (New York: Oxford University Press, 1999), p. 723.
2. As a beginning: "But each of the carefully, even painstakingly, crafted films Kubrick directed betrays the consistent themes and concerns of an acute and pessimistic observer of the modern world and the twentieth century in particular." Geoffrey Cocks, James Diedrick, and Glenn Perusek, "Introduction: Deep Focus," in *Depth of Field: Stanley Kubrick, Film and the Uses of History*, ed. by Geoffrey Cocks, James Diedrick, and Glenn Perusek (Madison: The University of Wisconsin Press, 2006), p. 4.
3. Geoffrey Cocks, "Death by Typewriter: Stanley Kubrick, the Holocaust, and The Shining" in *Depth of Field: Stanley Kubrick: Film and the Uses of History*, ed. by Geoffrey Cocks, James Diedrick, and Glenn Perusek, p. 197.

4. Julian Rice, *Kubrick's Hope: Discovering Optimism from 2001 to Eyes Wide Shut* (Lanham: The Scarecrow Press, 2008), p. 14.
5. Glenn Perusek, "Kubrick's Armies: Strategy, Hierarchy, and Motive in the War Films of Stanley Kubrick," in *Depth of Field: Stanley Kubrick, Film and the Uses of History*, ed. by Geoffrey Cocks, James Diedrick, and Glenn Perusek, pp. 87–88.
6. Kubrick had wanted a happy ending with the soldiers saved but Kirk Douglas demanded the pessimistic ending of the novel, which was finally decided upon. It could be argued that if the men had been saved, Colonel Dax would not have emerged a hero since such a decision would have been taken well above his level. Andrew Kelly, "The brutality of military incompetence: 'Paths of Glory' (1957)," *Historical Journal of Film, Radio and Television*, 13.2 (1993), 215–27.
7. For an account of the star's political image, see Anthony Lane, "Kirk Douglas's Heroism, Large and Small," *The New Yorker*, February 6, 2020. https://www.newyorker.com/culture/postscript/kirk-douglass-heroism-large-and-small. Accessed June 18, 2020.
8. Linda Hutcheon, *A Theory of Parody: The Teachings of Twentieth Century Art Forms* (New York: Methuen, 1985), pp. 54–55.
9. Kubrick's full-blown satires among which *Dr Strangelove* is one lack these "sincere" segments, complementing the irony in the others. Group Captain Mandrake, the only honorable person in the film, is played by Peter Sellers who has a comic presence and even that is intended to be funny. It could also be argued that the demand for non-stop laughter weakens the impact.
10. Glenn Perusek, "Kubrick's Armies: Strategy, Hierarchy, and Motive in the War Films of Stanley Kubrick," in *Depth of Field: Stanley Kubrick, Film and the Uses of History*, ed. by Geoffrey Cocks, James Diedrick, and Glenn Perusek, pp. 92–93.
11. Perusek, p. 95.
12. "We're never going to get down to doing anything about the things that are really bad in the world until there is recognition within us of the darker side of our natures, the shadow side." Stanley Kubrick, quoted in Penelope Gilliatt, "Mankind on the Late, Late Show," *Observer* (London), September 6, 1987.
13. "On his first day on the set, Dan Richter, the actor who played Moonwatcher, was informed by (Arthur) Clarke that the ideas of Robert Ardrey were 'central to 2001,' and that his 1961 book, African Genesis, 'caused quite a stir' over its assertion that 'we evolved from our ape ancestors because we learned to kill.'" Julian Rice, *Kubrick's Hope: Discovering Optimism from 2001 to Eyes Wide Shut*, p. 27.
14. Steven Spielberg, "Foreword," in *Stanley Kubrick: A Life in Pictures*, ed. by Christiane Kubrick (New York: Little, Brown, 2002), p. 10.
15. The Overman is the ideal superior man of the future who could rise above conventional Christian morality and otherworldliness to create and impose his own values, originally described by Nietzsche in *Thus Spake Zarathustra* (1883–85). "I love those who do not first seek beyond the stars for reasons to go down and to be sacrifices: but who sacrifice themselves to the earth, that the earth may one day belong to the Superman," https://www.gutenberg.org/files/1998/1998-h/1998-h.htm. Accessed January 29, 2025.

16. Mark Crispin Miller, "2001: A Cold Descent," in *Depth of Field: Stanley Kubrick, Film and the Uses of History*, ed. by Geoffrey Cocks, James Diedrick, and Glenn Perusek, p. 140.
17. Geoffrey Cocks, "Death by Typewriter: Stanley Kubrick, the Holocaust, and *The Shining*," in *Depth of Field: Stanley Kubrick, Film and the Uses of History*, ed. by Geoffrey Cocks, James Diedrick, and Glenn Perusek, p. 186.
18. Julian Rice, "Projecting Light: *The Shining*," in *Kubrick's Hope: Discovering Optimism from 2001 to Eyes Wide Shut*, pp. 115–53.
19. "Over the years, Stephen King has made his dissatisfaction with Stanley Kubrick's adaptation of his novel 'The Shining' quite clear. He has said Wendy is 'one of the most misogynistic characters ever put on film' and has described the director's film as being 'very cold.'" Kevin Jagernauth, "Stephen King Says Stanley Kubrick's *The Shining* is 'Like A Big, Beautiful Cadillac With No Engine Inside It,'" *IndieWire*, February 3, 2016. https://www.indiewire.com/2016/02/stephen-king-says-stanley-kubricks-the-shining-is-like-a-big-beautiful-cadillac-with-no-engine-inside-it-83995/. Accessed June 23, 2020.
20. It has been noted that the film "is less about ghosts and demonic possession than it is about the murderous system of economic exploitation which has sustained this country since, like the Overlook Hotel, it was built upon an Indian burial ground." David A. Cook, "American Horror: The Shining," *Literature/Film Quarterly*, 12.1 (1984), 2.
21. These is a "supernatural" sequence in which a mysterious ball comes rolling along as Danny plays with his toys and enters the circle, as if he gives emphasis to this, since it is not explained. The maze in the film is only a trap unlike in the book where the animals in sculpted from its foliage come alive.
22. R Barton Palmer, "*The Shining* and Anti-Nostalgia: Postmodern Notions of History," in *The Philosophy of Stanley Kubrick*, ed. by Jerold J Abrams (Lexington: The University Press of Kentucky, 2007), p. 216.
23. "We decided that somebody obviously had to die because it was a horror film. I thought it should be Danny, the child, but Kubrick was much more tender hearted than I and said, 'Oh, no, impossible.'" Diane Johnson, interview by Donald Williams, quoted in Julian Rice, "Projecting Light: *The Shining*," in *Kubrick's Hope: Discovering Optimism from 2001 to Eyes Wide Shut*, pp. 117–18.
24. Jonathan Rosenbaum, "In Dreams Begin Responsibilities," in *Depth of Field: Stanley Kubrick, Film and the Uses of History*, ed. by Geoffrey Cocks, James Diedrick, and Glenn Perusek, p. 246.
25. Michiko Kakutani, "A Connoisseur of Cool Tries to Raise the Temperature," *New York Times*, July 18, 1999, p. 22.
26. "Stanley Kubrick's *Eyes Wide Shut* (1999) is an existential film about human nature, sexuality, marital fidelity, and the nature and significance of choice. It is one of Kubrick's most optimistic films. The rituals and banalities of life that often exercise a deadening, soporific effect on the main characters in Kubrick's other films function here in a positive way, providing opportunities for awareness rather than obfuscation." Karen D Hoffman,

"Where the Rainbow Ends: *Eyes Wide Shut*," in *The Philosophy of Stanley Kubrick*, ed. by Jerold J Abrams, p. 59.

27. Slavoj Zizek, "*Eyes Wide Shut* and the Lacanian Real," The Kubrick Site. http://www.visual-memory.co.uk/amk/doc/0107.html. Accessed October 10, 2012.
28. Jeremiah Kipp, Eyes Wide Shut, *Slant Magazine*, November 20, 2007. http://www.slantmagazine.com/film/review/eyes-wide-shut. Accessed November 24, 2024

3

RW Fassbinder: The Original Sin

Rainer Werner Fassbinder has secured a place for himself in film history but there is little agreement on what are his best films; the same film (for instance *Katzelmacher*, 1969) features both on "best film" and "worst film" lists. This is unusual and if one were to consider his contemporaries like Werner Herzog, there would be general agreement that *Aguirre, the Wrath of God* (1972) is among his most important films. Fassbinder spoke about himself and his childhood at some length but there have been doubts expressed about the reliability of what he said.[1] There is not even agreement between critics on what issues each film deals with, and the same film is often described differently by various critics. Fassbinder was and remains an enigma and this is an effort to partly unravel it. He was extremely prolific and, when he died from an overdose of cocaine in 1982, he was at the height of his fame, having made within a short span of seventeen years an astounding 43 films.

Obviously, given these facts, no exploration of his work can be complete or even thorough. Films will need to be chosen almost subjectively for examination, an admission to be made right at the beginning. This is a book focused on irony and satire in cinema but where satire is usually quite obvious in that the encoder's mocking intent is grasped by the decoder, the satirical edge in Fassbinder is often far from that and needs to be deciphered. Fassbinder was the leading light of the movement that came to be called New German Cinema, which emerged around a decade after the French New Wave, i.e. around 1970. But while it may not be necessary to first understand the nitty-gritties of the movement, it would be helpful to know where German cinema was placed before, during, and after the War since Fassbinder came out of that ethos.

A brief history of German cinema up to 1970

Much of German film history is synonymous with the history of the Universum-Film Aktiengesellschaft or UFA, once the largest and most exciting film company in Europe. Ironically, UFA began and also finally ended as an instrument of propaganda. By 1917, the War Ministry in Berlin had begun to believe that the Hohenzollern dynasty was becoming unpopular and, in a meeting convened on May 25, 1917, it was decided that the Kaiser and his family should be brought into closer contact with the people through an information campaign. A reason for initiating the founding of UFA was also the moral turpitude once detected in German cinema,[2] although by 1913 the potential film as art had been grasped and an auteur film *The Student of Prague* made.

In March 1927, UFA came to be headed by Alfred Hugenberg, who was then well known for his right-wing affiliations. Hugenberg initiated a restructuring exercise that left the Deutsche Bank with losses but also gave him absolute control over its functioning and the Nazis eventually gained control. The subject of cinema under National Socialism can form the basis of an entire volume but for our purposes Nazi propaganda films were extraordinarily effective. The best known of the filmmakers to churn out films for National Socialism was the documentarist Leni Riefenstahl but her films like *Triumph of the Will* (1935) and *Olympia* (1938) are exceptions rather than the rule. Hans Steinhoff (*Ohm Krüger*, 1941) and Veit Harlan (*Jew Suess*, 1940) were crude emotional manipulators who made melodramas although their films were blatantly propagandist.

With the occupation of Germany after the war, the allies began the process of de-Nazification in right earnest and cinema could not but become an important cultural weapon. An Information Control Division (ICD) with a Motion Picture Branch (MPB) was set up in the US zone and the task of the ICD was to switch over from war-time propaganda to civilian reconstruction, "to assist in the reestablishment of a free German media on a sound basis and reorient the German spirit."[3] Films began to be screened regularly in the US zone in late July 1945 when twenty theatres were permitted to reopen. An urgent supply of American films was sought to convey to the people an "understanding of American life and democratic institutions."[4] The content of the films was strictly monitored and anything that might have shown America in a negative light—*The Grapes of Wrath* (1940) and even *Gone with the Wind* (1939)—was kept out, the latter apparently because of its portrayal of "objectionable Negro incidents."[5]

A crucial occurrence that was to have a lasting effect on the future of German cinema was the founding of film clubs by French cultural officers in the years of occupations. The screenings were attended largely by university students, teachers, professionals, and members of the educated elite and the films screened were early German classics or those regarded as controversial: Roberto Rossellini's films from post-war Italy, left-wing German films from the Weimar period. The repression in the air of the times can be gauged from the fact that when the Munich Film Club seceded from the national association, paranoid state officials openly expressed suspicion that this was motivated by a desire to secure the exhibition of propaganda films from the East Zone. By 1956, a group of the more radical younger members seceded from the film club movement because of the conservative tendencies exhibited by it. Enno Patalas, a doctoral student in Munster, and some of the others launched a left-wing film journal *Filmkritik*, which became the mouthpiece of the secessionists. The group drew upon the cultural criticism of the Frankfurt School, from writers like Walter Benjamin and Siegfried Kracauer. The approach of this group was deeply ideological and their initiative was perhaps as much responsible for New German Cinema of the 1970s as Andre Bazin and the *Cahiers du Cinema* had been for the French New Wave in the 1960s. German cinema had struggled for nearly 25 years in the shadows before it could emerge again.

Fassbinder's films

Rainer Werner Fassbinder had a deprived childhood and autobiographical elements abound in many of his films.[6] *I Only Want You to Love Me* (1975), based on an interview with a murderer, is about the unloved child who tries to buy love from his family; *The Merchant of the Four Seasons* (1971) about a vegetable peddler and is based on a favorite uncle who got no support from his family when he set himself up selling fruits and vegetables. Another aspect was Fassbinder's bisexuality; he gave his male friends female nicknames and called himself Mary, and this finds reflection in *In a Year of 13 Moons* (1978). His father, a doctor, was struck off the medical register either for drunkenness or illegal abortions, and he then bought a few dilapidated lodging houses, putting partitions in rooms and rented them out to immigrant workers; Rainer was sent out to collect rent from them. The immigrant worker features in *Katzelmacher* (1969) and *Fear Eats the Soul* (1974). There might be a case for examining his films in such terms but, regardless of the "personal" content of his films, they are often savage in their irony, implying a less-than-personal approach that might be dedicated to "feeling."

Fassbinder began his career making films based on his own plays, minimalist works with the flavor of theatre but nonetheless cinematic and under the influence of Jean-Luc Godard; it would be easier to get a purchase on his work by beginning with this category, represented by his first films *Love Is Colder than Death* (1969) and *Katzelmacher* (1969). Other categories would be his melodramas involving the working class inspired by the films of Douglas Sirk (like *Fear Eats the Soul*, 1974) and his melodramas set in the economic miracle (*The Marriage of Maria Braun*, 1978), and his adaptations of literature (*Despair*, 1978). The dividing line between categories is not always clear and could be contested; films of most categories, for instance, include gay relationships. Fassbinder worked in too many categories for a chronology to be advantageous and it seems preferable to first deal with some groupings through examples.

Early minimalism

Before embarking on a study of Fassbinder, it must be said that he has been subjected to an exhaustive study by Thomas Elsaesser and his approach is illustrated by the following passage about *Katzelmacher* (1969), one of the films I will be looking at in this section.

> A good example is KATZELMACHER (1969), Fassbinder's second film, which became not only his first claim to fame and author status, but also gave him—via different prizes and subsidies—valuable operating capital for years to come. KATZELMACHER was liked, uniting the avantgarde critics (impressed by the disciplined formal experiments) and those who expected from the cinema a social message, for it commended itself by a prescient analysis of the Gastarbeiter problem, in addition to satirizing German petit-bourgeois ambitions and Bavarian complacencies. KATZELMACHER's avantgarde status was also based on the impression that Fassbinder was following Bert Brecht's anti-illusionist precepts. Organized into short tableau scenes, the images are curiously frozen and inert, doubly deprived of depth by Fassbinder's head-on shots and static camera set-ups. When the camera does move, as in the lateral reverse tracking shots of the couples walking past a block of flats, a deliberately shallow space plays off the proscenium-stage against the illusionism of cinema. And yet, there is little evidence that a "model" of social conflict is intended, in the manner of Brecht's learning plays. Nor is KATZELMACHER a filmed play, despite the fact that it derives from Fassbinder's theatre-piece of the same title.[7]

My own scrutiny will try to be more closely textual than Elsaesser's—in keeping with the rest of the book where the narrative gets importance—with the expectation that it could still add something to what has been already said about his films. Irony and satire are also located largely in the narrative.

Fassbinder began his career studying acting in Munich between 1964 and 1866 with the intention of going on to the stage. He made several 8 mm films, write a couple of plays but was denied a seat in the Berlin Film School. He returned to Munich, continuing to write plays and made two story films of which the first *The City Tramp* (1966) was apparently inspired by Eric Rohmer's short film *The Sign of the Lion* (1959). He met Hanna Schygulla who was to star in a large number of his films and cast her in his second short film *The Little Chaos* (1967), which he said was a bit like Godard, and the influence of the French New Wave may have indirectly owed to New German cinema's origins in the film clubs started by the French after the War. Although shy, Fassbinder was a natural leader and joined a Munich theatre group "Action Theatre." He came to head it in a few months and staged his own play *Katzelmacher* successfully. But the group broke up due to rivalries and he founded another group Anti-Theatre and many of the actors from Action Theatre—Peer Rabe, Harry Baer, Kurt Raab, Irm Hermann, and Hanna Schygulla—were happy to allow Fassbinder to be their director. His first feature film *Love Is Colder than Death* (1969) shows theatrical stylization but it is also highly cinematic, in a minimalistic way, and reminiscent of Godard's *Vivre sa Vie* (1962).

Love Is Colder than Death (1969)

This is a gangster film that resembles a French New Wave version of an American gangster film although it is a striking debut film on its own. The film begins with a group of men waiting for something; Franz (Fassbinder) is reading a newspaper and another man asks him for a cigarette. But when it is not forthcoming, he snatches the newspaper and throws it on the ground. Franz gets up lazily, goes up to this man, and punches him into unconsciousness and resumes reading the newspaper and smoking. Franz is next being interviewed blindfolded and offered a job with the "Syndicate," well paid if he does what he is told, but Franz refuses, only to be pistol-whipped. The filmmaking style is striking, there is very little movement and the scenes are almost like tableaux. Much of the violence happens off-screen and unconscious men are deposited on the floor.

What emerges is that the Syndicate (a bare-chested Black man speaking English, indicating American connections, with a pistol in a holster) is recruiting personnel and these toughs are potential candidates. Franz is a pimp and has been in jail; he is from Munich and lives with a prostitute Joanna (Hanna Schygulla) whom he exploits. At the recruitment, he meets handsome Bruno (Ulli Lomel), from a small town, is drawn to him, and invites him to Munich, giving him his address. Bruno comes to Munich, locates Franz and Joanna, and the three share lodgings. Franz is hiding since a Turk accused him of having killed his brother and wants revenge.

There is a homo-erotic charge in the relationship between Franz and Bruno and Franz wants Joanna to sleep with Bruno, but she is not inclined to because of her love for Franz, behaving more as a wife than a prostitute might with a pimp. Franz is made angry by this fact and slaps her for making fun of his friend. The trio visits a mall, steals dark glasses, and then buys firearms. Bruno is a psychopath and kills the gun dealer, who is also a cobbler. Later the trio kills the Turk in a café and Bruno also kills the waitress (named Erika Rohmer). Franz is arrested and questioned for a day and Joanna gives herself to Bruno, meanwhile. There is a moment of conviviality between Bruno and Joanna after this, when they saunter through a mall doing a bit of shoplifting together, uplifting music on the soundtrack. The three now plan to rob a bank but we come to know that Bruno is with the Syndicate and is instructed to kill Joanna. But Joanna has also made contact with the police; and the long and short of it is that it is Bruno who is killed. Joanna admits this to Franz who responds by calling her a "whore," although perfunctorily.

The acting style used in the film is the very opposite of that in the French New Wave and the characters are all inarticulate. It is not that they do not speak out thoughts as much as that they have none. These are the lumpenproletariat and, if they are criminals, they are of a breed different from Michel Poiccard in Godard's *Breathless* (1960), still an existential hero who unwittingly raises philosophical questions. Fassbinder as Franz has a particular gesture worth noting, which is to stare down at the floor, his crown pointing at the audience and Hanna Schygulla is also made to maintain the blank countenance of someone not accustomed to thought. I would see this as a deliberate parody of the French New Wave; Fassbinder's protagonists are essentially unthinking louts. To substantiate the claim is Bruno's train ride to Munich when he is seated opposite a woman who is drawn sexually to him, so much so that when he takes money out of her purse, she does not object. The sequence is filmed with the window at the center of the frame, with the cityscape rushing by. One recognizes in the sequence an allusion to Godard's *La Chinoise* (1967), where a young woman discusses revolutionary politics with philosopher Francis Jeanson on a train as the landscape rushes by. In Fassbinder's film, Bruno asks the woman if she is thinking of sex and she replies that she is thinking of "the revolution!"

This sense of his protagonists being "stupid" occurs often in Fassbinder's films as I will try to demonstrate. But "stupidity" means a lack of "comprehension" and the best way to define that incomprehension is in relation to the "larger picture." In *Love Is Colder than Death*, for instance, the big picture is represented by the "Syndicate" that Franz does not inquire into; he does not even gather that Bruno is working for it and he is living from moment to moment off Joana. But this incomprehension is not Franz's alone; it is a malaise that affects most people, it seems. Bruno kills without reason, although that might get them into trouble, and takes stupid risks with a bank robbery scheduled. The trio has just killed a policeman for no reason but the police do not bring it up when questioning Franz. A client of Joanna's is so eager at the prospect of being with her that it takes him a while to notice Franz and Bruno in her room, armed with guns; Joanna needs to direct his gaze toward them by turning his head away from her. This incomprehension is not a Pinteresque reflection on the large scheme being incomprehensible[8] but of small people too caught up in their petty desires to reflect upon the larger issues that directly affect them. This is important because Fassbinder's social range in dealing with Germany was immense[9] and he sees other people as "thoughtless" in other films as well. He saw his country in terms of its people and one does not get a sense of the landscape or vistas that one does with Herzog or Wim Wenders when dealing with Germany.[10]

Katzelmacher (1969)

If *Love Is Colder than Death* works partly as a parody (of Godard), *Katzelmacher* is Menippean satire. This film, which is also based on one of Fassbinder's plays, had been staged earlier by his group Anti-Theatre. This film is about a group of slackers who are unemployed or partly employed and hang about beside an apartment block in Munich. Erich (Hans Hirschmuller) and Marie (Hanna Schygulla) are together, as are Paul (Rudolf Waldemar Brem) and Helga (Lilith Ungerer). Rosie (Elga Sorbas) is unattached as are Franz (Harry Baer) and Gunda (Dorres Mattes). The men are together but the women are either preoccupied with retaining their men or are poaching on a friend's turf. Franz prides himself on having money and pays Rosie for sex, even bragging to the others about it later. Paul is bisexual and has a male lover that the others do not know about. Elizabeth (Irm Hermann) does not socialize with this group but lives with a man in one of the apartments. She has some money and is therefore ill-mannered with the man who depends on her financially. The members of the group, like the ones in *Love Is Colder than Death*, are hopelessly inarticulate and their conversation focuses largely on rumors about each other told to the others when the object of the rumor is absent. They bitch about each other on weekdays but on Sundays they link arms, with uplifting piano music on the soundtrack, although even then, their rivalry is apparent:

MARIE: My man cares about me.
HELGA: So does mine.
MARIE: Erich gives me a lot of love.
HELGA: Paul also.

Into this milieu comes Jorgos (Fassbinder), an immigrant worker from Greece who takes up lodgings with Elizabeth. Since the room he is to occupy is being painted, he shares a room with Elizabeth's man. The next day the group accosts the man to ask about the Greek and learn that his penis is larger than those of Germans, and this information is met with a minute of silence. Later Gunda runs into the Greek and asks Jorgos about Germany and German women and the subject comes around to "ficky-fick." The message is duly transformed and it finally becomes "Jorgos raped Gunda."

Only Maria likes Jorgos because he looks straight at her while talking and she leaves Erich, who has been violent. Jorgos does not speak German well but uses his ignorance of it prudently—protesting his perplexity when Maria asks him about his wife in Greece. Soon Jorgos and Maria are seen together, leading to more hostility from the group. Elizabeth has been overcharging Jorgos and, although she is disliked, this has their approval. Finally, matters come to a head and Jorgos is assaulted by the men including Elizabeth's man who cannot cogently explain his conduct to her. The men conclude that he will vacate and leave, but Jorgos is ready for more "boom-boom"; Elizabeth however gains because she raises the rent.

I described the protagonists of *Love Is Colder than Death* as "stupid" and the same is true for the group in *Katzelmacher*. Jorgos is a Gastarbeiter (guest worker) invited to work in Germany and "katzelmacher" is the vulgar term used for Gastarbeiters, comparing their sexual proclivities to those of cats. This becomes an issue for the men and in one segment pertaining to the period when they are still friendly with Jorgos, they sit at a table with him, pass comments about his sexual apparatus and propose his castration, touching glasses with him amiably now and then. We cannot be sure that Jorgos does not follow their talk since he understands more than he acknowledges. Fassbinder does not make Jorgos a sad victim of xenophobia but someone canny who is aware of his situation. He also harbors racist feelings toward a Turkish Gastarbeiter in his place of work. The notion of "guest workers" was a state decision based on industry's need for workers and Gastarbeiters contributed to the economy in both Germany and their home countries. They were perhaps more valuable to resurgent Germany than the quasi-lumpen types represented by the group. Yet, being Germans, they believe a bit of violence can drive Jorgos back to Greece. The conversation also reveals their ignorance of a country also in Europe, speculating about sexual habits there as though it is on another continent.

Two aspects of his films need to be discussed here and the first is his reliance on the same troupe of actors in film after film. Ingmar Bergman also did this but Fassbinder, it could be argued, was closer to the British *Carry On* comedy series produced by Peter Rogers and directed by Gerald Thomas between 1958 and 1978. The attraction of that series was that the same actors played a variety of roles with subjects ranging from Julius Caesar and Cleopatra (*Carry on Cleo*, 1964) to Britain's colonial doings (*Carry on Up the Khyber*, 1968) or even contemporary stories, but in the *same way* each time. For instance, Julius Caesar in one film would be played in the same way by the same actor in another film where he would be a "Nabob" or in a third one where he would be the resident surgeon in a London hospital. The effect is comedy, perhaps satirizing failed artifice, but with Fassbinder—although the actors are immediately recognizable from their other roles—it is slightly different though what is emphasized is artifice.

Fassbinder used the same actors but rarely does any of them don a disguise to be someone else.[11] The only actor who tries to actually play at being someone else each time is Fassbinder, although he restricts himself to "impersonations," getting the gestures right without "inhabiting" the roles. The strategy is effective because the actors underplay and allow audiences to work at seeing them as different people, or do over-the-top representations drawing attention to it as performance. By this method (a consequence of the actors being from his theatre troupe), he came up with an authorial approach, asserting that aspects of the same complex reality could be represented with common equipment, affirming their unity and also the underlying artifice in the exercise. This is something he drew from theatre where verisimilitude is not insisted upon. The same props become diverse things and we accept their amended meanings.

The second aspect is Fassbinder's extensive use of nudity, often with the same actors exhibiting themselves. I would argue that it is not "nudity" but "nakedness" in Fassbinder's films,[12] i.e. disguise "dropping off," given the awkwardness in evidence. Often, this gets further emphasis when only one of the persons is naked while the other is fully clothed, as in *Love Is Colder than Death*. Apart from their meaning in the narrative, Fassbinder is not presenting "characters" in these scenes as much as the actors without disguise, demanding that we still accept them, the way Pablo Picasso demanded that we accept a bicycle seat and handlebars as a "bull's head." Nakedness in a role becomes a kind of identification: "being" instead of seeming.

The working-class melodramas

Fassbinder favored the melodrama as a form although a deeper look at his films shows that they are far from what we normally call melodramas. Broadly speaking, melodrama is seen to indulge in strong emotionalism, moral polarization, and schematization; it portrays extreme situations and actions, overt villainy, the persecution of the good and the final reward of virtue, inflated or extravagant expression, abrupt changes in fortune, and dark plotting and suspense.[13] The origins of western melodrama have been located within the context of the French Revolution and its aftermath when the traditional sacred and its representative institutions (Church and Monarch) lost their centrality. The agent of morality became the Republic and, where narratives had depended on the traditionally sacred institutions for mediation, a substitute in the shape of a "moral occult" was incorporated into narratives, a metaphysical system without affiliations to the Church or monarchy to provide resolutions.[14] Melodrama produces effects that are the opposite of what was said about Luis Buñuel's realistic films in that there is a reward for virtue and punishment for wickedness and the filmmaker plays God in dispensing them. Two of his best-known melodramas deal with the plights of working-class protagonists but Fassbinder is not really doing what one might expect by "playing God."

Fox and His Friends (1975)

This film begins in the fairground where Franz Biberkopf (Fassbinder) is being presented as "Fox the talking head" when the impresario running the show, his boyfriend Klaus, is arrested for tax fraud and the show closes down. His drunken sister Hedwig is of no help and he has run out of money, not even enough for a lottery ticket. In a public toilet, he meets Max (Karlheinz Bohm), an antique dealer who helps him get a ten-mark lottery ticket through subterfuge by cheating a gay florist. A few weeks later, after he has won half a million marks in the lottery Max introduces him to his friends and Franz takes a lover Eugen (Peter Chatel). Franz is from the working class and all these friends are cultivated. On Max's advice, Franz invests in Eugen's father's ailing business (a printing press), lends them 100,000 marks, buys an apartment in Eugen's name, and furnishes it expensively with furniture bought from Max. He works at the printing press hauling material—since he is of the working class, anyway—while Eugen sits at a desk in his office. This is ironic since having become wealthy Franz should have sat at the desk! He loans his former boyfriend Klaus 30,000 marks when the latter comes out of jail.

Franz is incurably generous with his money and takes Eugen on a holiday to Marrakech in Morocco, where they try to pick up a Moroccan male prostitute, who is not allowed into their hotel, which keeps out Arabs. In short, Franz makes bad, trusting decisions and loses all his money. He does not get his loaned money back from Eugen's family and also has his working-class manners constantly made fun of. Whatever can go wrong does go wrong and the film ends with his death from an overdose of Valium, his last remaining money grabbed by two passing schoolboys.

Fox and His Friends is highly regarded by critics and they take Franz's victimhood very seriously as social criticism. Here is Elsaesser noting the complaints of critics against the film:

> When a film puts sex and class on a parallel track, as does FOX AND HIS FRIENDS, the story of a proletarian hero who is both sexually exploited and economically ruined by his bourgeois lover, while his lack of education, his peasant manners and naive adulation of middle-class values make him a laughing stock for the snobs he so much admires, the narrative can easily appear didactic, for it seems to lack the dimension of poetic justice or ironic reversal that gives the stages of the Fassbinder

> plot usually such a grimly satisfying shape [...]. Yet, rather than trying to find a way to free themselves. by showing his protagonists coming to a raised consciousness, breaking out of the family in order to become militants, which was the demand of a certain "political film practice, and of 'progressive-realist' film theory, Fassbinder appears to let his characters rot in their despair."[15]

But my own reading of the film is that Fassbinder is not empathetic to Franz but actually mocking his naiveté. Franz is cunning enough to steal ten marks from a florist but cannot protect his half a million marks. His innocent countenance is constantly juxtaposed with Eugen or Max's conniving one and we are, at every moment, made uneasy by the fact that Franz is being ruthlessly cheated by well-bred scoundrels. If there are doubts about this over-the-top representation of an innocent protagonist with a heart of gold done in by false friends being facetious, the two children stealing the last of Franz's money from his corpse should clear them.

Fassbinder's film so uses every means to drive home its tendentious "didactic" message that one is led to look elsewhere for the film's value. He himself seemed to pronounce his melodramas as "Brechtian" in some sense, i.e. make his audiences reflect upon what he is showing.[16] This could be true but it is certainly not the social message in Franz's victimization that we should reflect upon, since the level of Fassbinder's seriousness is in doubt; there is far too much hyperbole in it.

I would say that the most telling scene in the film achieves just the opposite of what the rest of the film does; Franz is in Eugen's flat looking around but with a new air of cockiness as someone equal in status and fully entitled to be present there, doing things like wearing his shoes in bed and disinclined to hide in the bathroom when Phillip, a former lover of Eugen, turns up. Franz's trousers, Phillip notices, have the smell of a public lavatory but Franz, with his newly acquired wealth, does not take "class" into consideration; only wealth, that he believes places him on par with his lover. This means that Franz, despite the pathos of his role, is on par with the obtuse protagonists of the earlier films, uncomprehending of the social milieu in which he lives. Fassbinder is essentially subverting the thick melodrama in his film through this ironic element. Mocking the victim for his/her stupidity—by overemphasizing his pathos and slipping in a blistering indication that people should take care of their own interests—is clearly unusual.

Then there is the full, male frontal nudity in the film, Franz, Max, Eugen, and the others exposed time and again to the audience. The first meaning in presenting a self-serving gay community is that gay love is no less exploitative than any other kind and involves money as any other. But more subtle than that is Fassbinder's strategy of filling his film with naked men; I would argue that the nakedness creates the illusion that people are equal. Its satirical aside is that Franz Bieberkopf is confident initially not only because of his newly acquired wealth but also on account of the size of his penis, larger by his own claim than Eugen's. But once people start dressing up, class differences start emerging when Franz dresses in a leather jacket while the others display "cultivation." Visiting the tailor or the clothier is a key happening in the film when people dress according to class. The difference between a person's naked persona and his clothed self says much about the class structure in the gay society portrayed in the film.

Fear Eats the Soul (1974)

This is a film made earlier to *Fox and His Friends* but also a melodrama cunningly subverted as *Fox* was. It is inspired by Douglas Sirk's *All that Heaven Allows* (1955) about a wealthy middle-aged widow (Jane Wyman) who falls in love with a younger working-class man (Rock Hudson) well below her station and the social ostracization she faces, partly from her own children. Fassbinder's film is a radical extension of it in that the woman is elderly and the man is a young African and there is much more conspicuous divergence between their respective bodies.[17]

In *Fear Eats the Soul*, an elderly Munich cleaning woman Emmi Kurowski (Brigitte Mira) meets Ali (El Hedi ben Salem) a Gastarbeiter from Morocco, 30 years her junior, in a bar and is drawn to him when he asks her to dance with him. He accompanies her home, stays the night at her invitation, and shares her bed. Later, when the landlord calls to tell her that she cannot take in lodgers, the overwhelmed Emmi lets slip that she is marrying Ali. Ali hears this, immediately affirms that it is a good idea and the two are therefore married. Emmi is familiar with the prejudices against foreigners since her first husband was a Pole and her marriage to him was met with resistance from her father, a member of the Nazi party, which she also joined. She now faces ostracization from her own children, colleagues, friends, neighbors, and the local shopkeeper. Fassbinder plays up this aspect as in *Fox* and also their hypocritical natures when they return to her to make friends again at a time when they need favors from her.

All this is standard melodrama and Fassbinder stages it expertly, people staring at them in all sorts of places, standing in groups sometimes, gazing out at the couple, sitting at a table. This is not realistic but akin to theatrical stylization with evident exaggeration.[18] But *Fear Eats the Soul* would not have amounted to a masterwork if Fassbinder had not slid in another covert aspect into the film. Emmi has a roomy basement and a neighbor wants to use it for her son's possessions and Emmi sees it fit to offer Ali's services. Ali helps but one catches the subtle change that has been introduced into the couple's relationship by her offer: her husband's physical labor and also announcing his strength to someone of her own station that she wishes to get on with. Emi and Ali quarrel a while later because she will not make couscous but this offer is the primary, unspoken reason. Ali is upset, gets drunk and returns late, and she leaves him to lie in the corridor. Later she praises Ali for actually showering every day and also inviting friends to feel his muscles. Evidently, this leads to a feeling of rancor in Ali, which is never made explicit, and he takes to sex with Barbara (Barbara Valentin), the bar girl.

When at work, another foreigner, a woman from Herzegovina arrives Emmi behaves the way the others behaved with her, moving away with them from the new woman to exchange confidences. Emmi is a sad and lonely person but as the film progresses we begin to realize she shares most of the prejudices that she has herself been the victim of. Decades earlier, we recollect, she had married a Pole and joined the Nazi party, which devastated Poland. The German-Polish conflict is the subject of another celebrated film from New German Cinema, Volker Schlondorff's *The Tin Drum* (1979).

Emmie, it appears, still admires Hitler and takes Ali to a restaurant that the later Fuhrer once frequented. Not many critics may concur[19] but Emmi's is an ironic portrayal of someone incapable of comprehending the implications of her own socio-political choices. She is a logical continuation of Jorgos from *Katzelmacher* who shares the same prejudices that he was subjected to.[20] That there are some genuinely tender moments in the film involving her and Ali do not change this aspect of Emmi's stupidity since even stupid people are entitled to love. But when Ali is eventually ill with a gastric ulcer that may be incurable, she voices determination to nurse him out of it; still, we in the audience, are not so confident of her ability to do so. Regardless of his seemingly empathetic approach, Fassbinder's political viewpoint arguably remains unchanged from *Katzelmacher*.

Melodramas set in post-war Germany and the economic miracle

Two of Fassbinder's melodramas dealing with working-class protagonists were just discussed and neither of them takes up the economic miracle as its setting. There is also Fassbinder's ironic view of the economic miracle in the person of Anton Saitz in *In a Year with 13 Moons* (1978). Saitz is a Jewish survivor of the Holocaust who set up a brothel in post-war Germany with "Nazi efficiency." The two films to be discussed here provide us with more of the same, what seems to be melodrama being turned around; but Fassbinder's subversive touch as in the earlier melodramas, raising different questions from those suggested by the plot, is less in evidence here.

The Marriage of Maria Braun (1978)

The film begins in 1943 and Maria (Hanna Schygulla) is marrying Herman Braun (Klaus Lowitsch) during an Allied bombing. The registry office where the marriage is being officialized is bombed and the certificate is stamped with the official lying down in a shelter. This is obviously hyperbole and the rest of the War is deliberately pictured as overstatement and cliché. Maria and Hermann are together only for a night and half a day; Hermann does not return after the War and Maria visits the railway station in vain with his name on a placard. The deliberately clichéd tone Fassbinder chooses can be caught through the detail that the placard falls under a train, to be crushed.

Maria now finds a job as a hostess in a bar and meets Bill, a black American soldier. American soldiers are in demand because of their access to cigarettes; in a café Maria is witness to an American soldier speaking slightingly of her group and her strong response has him apologizing, with a packet of cigarettes. The sexual relationship between her and Bill is portrayed strikingly through the contrast between the colors of their sweating skins but, when the two are undressing each other, Hermann Braun returns. Neither Maria nor Bill notices him and the interminable segment showing Hermann watching the two being intimate reeks of cliché until Maria freezes when she sees him. But Hermann rushes first to a packet of cigarettes on the table—before he and Bill get into an altercation, a fight that Maria interrupts by smashing Bill's head with a liquor bottle. Maria is tried for murder but Hermann Braun takes responsibility and goes to jail in her place.

The opening of the film had machine gun fire on the soundtrack and the economic resurgence of Germany has similar-sounding demolition hammers on the soundtrack, buildings being pulled down and part of the same activity that Anton Saitz was involved in and, as also in *In a Year of 13 Moons* football is a sign of German resurgence. Maria now accidentally meets a German industrialist Karl Oswald (Ivan Desny) on a train, impresses him by subduing an American GI who was turning abusive, and becomes Oswald's secretary and his lover. She lives with her mother, grandfather, sister, and brother-in-law and we see the family moving up the social scale.

Maria loves Hermann and tells him about Oswald and Oswald realizes it is her husband she loves. The film proceeds on these lines, with amusing asides like Maria's favored brother-in-law being a union leader who she is to officially "fight." The film is listed among one of Fassbinder's greatest by critics and there is little doubt that that it is strikingly made. Still, the long section in which Maria has to deal with two men, both good people who love her, does not aim for complexity. It is, rather, only holding out for something dramatic to happen. Eventually, Oswald dies and Hermann returns to her from Canada—where Oswald had sent him as part of a deal—to share her new house but she dies with him in a gas explosion without ever finding true happiness.

The question one could ask here is why Maria cannot find happiness in a melodrama when she comes out of difficulties with her integrity not compromised. An answer could be suggested in the manner of their death—"gas." We know that victims in concentration camps died by gas—but differently from Maria Braun, to be sure. Fassbinder could be suggesting that she cannot find happiness because of the burden of the original sin of Nazism and the Holocaust that the German people must still bear, despite Germany's resurgence. Still, this is in the nature of a pun on the word "gas" and one cannot be too happy. But to western audiences, the film was an allegory of West Germany since it is peppered with references to former chancellors and West Germany's rearmament.[21]

The Marriage of Maria Braun has been written about by Elsaesser but the thrust of his approach is on the reason for its commercial success since it was Fassbinder's single completely uncontroversial film. Cited is its classical structure, its focus on a character played by a star, and, most importantly, it allowing German and non-German, male and female, mainstream and highbrow spectators to find a pleasurable emotional and intellectual point of entry. But judging from what Elsaesser writes, Fassbinder did not have much faith in the project:

> When one learns how MARIA BRAUN came to be made, one is struck by the unlikelihood of it existing at all. Or rather, one marvels at the many kinds of irony involved in the process, the role of accident and chance at every turn, seriously suggesting that the argument about the purposiveness of Fassbinder's life and work must be an illusion due to the historian's narrowing retrospective perspectivism. By most accounts, the making of MARIA BRAUN was one of Fassbinder's least happy experiences. Peter Berling even calls it "one of the decisive self-destructive episodes in Rainer's life." When the production was ready, Fassbinder no longer wanted to shoot the film, too preoccupied with the preparations for BERLIN ALEXANDERPLATZ to care about much else. Considering how important Hanna Schygulla proved to be for the film's success, it is ironic that originally Romy Schneider should have been cast for the part, before her demands, indecisiveness and heavy dependence on alcohol proved too irritating for the impatient director.[22]

Lola (1981)

The two films considered in this section mark a decline for Fassbinder. There was already evidence in *Maria Braun* that Fassbinder, having understood how the international festival system works, is using his considerable cinematic skills less to explore new content than to "play" it, since that film is like a Rorschach inkblot; anyone takes out of it as much as he or she puts in. Here is a section from a biography of Fassbinder that describes his last year 1982:

> In February, he had won the Golden Bear at the Berlin film festival with *Veronika Voss*, his previous production *Querelle*, he hoped, would win the Golden Palm at

> Cannes, and during the coming summer he would shoot his next film *I'm the Happiness of this World*, which the plan went would take the Golden Lion at Venice in September. The larger ambition was to win the Oscar for the best director in April 1983 and to make the cover of *Time*.[23]

With this information, his later films seem to emerge with different significance from his earlier ones and *Lola* is one of his final films. *Lola* is a story set in a small town Coburg in Bavaria teeming with corruption with Schuckert (Mario Adorf) the crooked property developer who controls everything. Schuckert also owns the local entertainment hub and brothel Villa Fink where Lola (Barbara Sukowa) is a singer. Into this town comes Von Bohm (Armin Mueller-Stahl) the building commissioner who is strict and honest and sets about trying to clean up the town. The film is based on Josef von Sternberg's 1930 classic *The Blue Angel* in which an authoritarian professor falls in love with and chases after a singer in a night-club who leads him to humiliation and death. Here, Von Bohm falls in love with Lola (as Marie Louise). The climax of the film is when Von Bohm understands Lola for what she is. While we begin by anticipating Von Bohm's success in cleaning up the town, the film ends with him marrying Lola, an ironic reversal with all-round happiness for the good and the corrupt.

The film is exquisitely shot by using heightened lighting with reds, blues dominating, which gives much of it the look of a staged performance; Fassbinder is apparently alluding to the film as artifice but *Lola* does not add much to our understanding except affirming the cynical truism that every man has his price. There is also the sense of the world not as it "is" but refracted through the radio[24] but this strategy has become mannered and does not yield insights. Fassbinder apparently believed that a good director could contrive an unsatisfactory ending that audiences would disbelieve in and that may be the strategy in both melodramas examined in this section.[25] Still, they do not have much to add about the economic miracle or about personal socio-political choices and conduct that the earlier films were rich in. It relies on the reversal of *The Blue Angel* plot, the happy ending becoming an ironic take on the economic miracle.

Lola is a film—also like *Lili Marleen* (1981)—that makes use of pop music and Fassbinder's use of pop hits is significant:

> "Typical for Fassbinder's use of popular hits is that they are not meant ironically, and nonetheless do not function as the illusion of harmony, as they do in other German musicals. The fabricated and self-deceiving emotionality of the pop songs is genuine, because the characters who sing or listen to this music, hang on to its promise for dear life. However, because Fassbinder at the same time shows how gagged, cheated or trapped these characters are in their environment, education and private traumas, he unmasks the pop songs as yet another lie by which they are betrayed.[26]

Determining a trajectory

As already set out, this study of Fassbinder's films avoids a chronology but, rather, looks at categories in his work. But since there are so many films and so many categories, it is difficult to do justice to his oeuvre; his trajectory as a filmmaker, his changing approach based on what we have already seen, also seems important. I have argued that "unawareness" on the part of people on their own socio-political positions and choices is something he satirizes in his early films. Fassbinder made feature films between 1969 (*Love Is Colder than Death*) and 1982 (*Querelle*), and I detected this same preoccupation till as late as *Fox and His Friends* (1975).

If a question to be asked at this juncture pertains to the nature of this unawareness, I propose that it is political unawareness, people remaining ignorant of the consequences of their choices. Emmi, I noted, was a member of the Nazi party in *Fear Eats the Soul*, perhaps a casual one, but she understood very little after her dalliance with Nazism and her goodness is still contaminated by racist notions. *Jail Bait* (1973) is a film about an ill-fated youthful romance in which the young man, at the instigation of the pregnant 14-year-old girl, murders her father for his opposition to their union. The father is a former Nazi sympathizer who would "rather gas a hundred thousand Jews than have his child done wrong." It would seem from these examples, although this is not conclusive, that it was the way the German people were led by Hitler and the Nazis that Fassbinder saw as the typical responses of people to politics—that the consequences of their political acts or choices would elude them. Of the films I have analyzed or merely invoked, only a few invoke Nazism explicitly but the motif still stands out.

But if there is a change in Fassbinder's approach between his early satires and later melodramas, here is in on the subject:

> I thought that if you brought people up against their own reality, they'd react against it. I don't think that anymore. I now think that the primary need is to satisfy the audience, and then deal with the political content [...] (using) the emotions generated to a particular end. It's a preliminary stage in a kind of political presentation. The main thing to be learned from American films was the need to meet the entertainment factor [...] I find the process beginning in Douglas Sirk's films or in a film like Hitchcock's *Suspicion*.[27]

In Hitchcock's *Suspicion* (1941), a guileless woman (Joan Fontaine) marries a handsome scoundrel (Cary Grant) and gradually comes to suspect that he is a murderer and she could be his next victim. By all logic, she should also have been killed, but Hitchcock puts in a fake ending indicating that her suspicions are unjustified. Fassbinder's implication could be that one could put in false endings with the expectation that the audience would reject them and catch on the truth about their world, and this is what he does in *Lola*. The next step, it would seem to me, is Fassbinder gradually coming to realize that the audience would believe the false ending instead of rejecting them and applauding him on that basis. The conclusion is speculative, but we may suppose from his ambitions (winning prizes and being on the cover of *Time*) that he gradually lost faith that his films would be understood politically. Fassbinder's film *Mother Kuster's Trip to Heaven* (1975) was criticized by the left, *Fox and His Friends* made him enemies among gays; in a *New York Times* obituary, Vincent Canby remarked that although Fassbinder considered himself a man of the left, it would not be surprising, after seeing *The Third Generation* (1979), if Fassbinder had eventually wound up with the right. We may take this to imply, for Fassbinder, a lack of faith in what would have seemed correct politics.

Interpreting Fassbinder's purpose

Reading the criticism written about some of Fassbinder's films, one does not get the sense that anything resembling an understanding of his political purpose was widely reached about them. It could hence be revealing to look at one of his later films, the adaptation of a literary classic since a literary classic at least furnishes us with a reference mark against which the film can be measured. How a filmmaker adapts it could tell us something about his or her artistic propensities and concerns. The film is *Despair* (1978), an English language film starring Dirk Bogarde and Andrea Ferreol, based on a novel by Vladimir Nabokov, with a sparkling script by British playwright Tom Stoppard. Fassbinder was evidently mounting an international production with big ambitions—perhaps the attainment of much wider acceptance.

The novel *Despair* (1936) is about Hermann Hermann whose schizoid tendencies lead him to undertake a preposterous plot to murder a man and pass him off as himself, collecting the insurance money and living happily ever after in Switzerland. The only difficulty is that the man he chooses bears no resemblance to him. The novel is written in the first person and is introspective although it works as a black comedy. Originally written in 1934 in Russian, it makes no references to the Nazis, which is the element Fassbinder introduces along with ornate indoor settings laden with mirrors. Fassbinder was evidently being obfuscating about his use of the mirrors:

> Nabokov's allusive, punning style, claims Fassbinder, finds its visual equivalent in mirrors and reflections: "Nabokov handles language as something that reflects, mirrors, and re-reflects," says Fassbinder. "If I use glass and reflections in this film, it's derived

> from Nabokov's linguistic structure [...] The significance of glass in this film is that it's transparent and yet it closes in on the character—and that's Nabokov's style."[28]

One wonders what to make of this since that is hardly clear and Fassbinder's collaborators had problems with the film; scriptwriter Tom Stoppard particularly disliked it:

> As Fassbinder describes them, the film's mise en scène and shooting style undermine the authority of screenwriter Tom Stoppard. That this is the first film whose screenplay Fassbinder didn't write or collaborate on goes some way towards explaining his insinuation that, in *Despair*, authorship resides in the domain of the visual. And there is more: Stoppard reportedly disliked the film, and was angry at Fassbinder's disregard of his script during the shooting and editing, complaining to Dirk Bogarde that "the crux of it is that I wanted to write the script because it was Nabokov's book, and Fassbinder wanted to film the story despite it's being Nabokov's book."[29]

Despair is a self-consciously arty film. In an early part, Fassbinder makes his protagonist Hermann Hermann sit back in an armchair and watch himself making love to his enormously stupid wife Lydia ("Intelligence would take the bloom off your carnality," he tells her). The ornamentation does not shed light on his "schizophrenia" and the value of the film lies elsewhere, in its treatment of the rise of Nazism in the early and mid-1930s, which does not engage Nabokov's novel at all.

There are several fleeting references to it in the film and one is initially tempted to dismiss it as detail to give it historical authenticity. In one sequence we see Brownshirts stoning a Jewish residence and in another, dogs are rounded up and taken to the pound. Hermann is a chocolate manufacturer and his factory produces dolls made of milk chocolate. During one of Hermann's visits to his factory, we are given a lingering close-up of these naked chocolate dolls lying together in a heap. What Fassbinder is suggesting is characteristically grotesque: the charred human bodies of Auschwitz and Birkenau,[30] the familiar images that immediately conjure up Nazism for us today.

Lydia has a cousin Ardalion, a painter of sorts, and as Hermann and he rummages through his studio looking for a previously painted portrait of Hermann. Instead, what they find is a still-life on a canvas with a Nazi Swastika scrawled behind it, perhaps by a child. The scrawled Swastika does not engage them. Here is a pertinent passage from Walter Benjamin (writing on Dadaism) that might shed some light on the sequence:

> The revolutionary strength of Dadaism lay in testing art for its authenticity. You made still-lifes out of tickets, spools of cotton, cigarette stubs, and mixed them with pictorial elements. You put a frame round the whole thing. And in this way you said to the public: look, your picture frame destroys time; the smallest authentic fragment

> of everyday life says more than painting. Just as a murderer's bloody fingerprint on a page says more than the words printed on it.[31]

The child's scrawl in *Despair* is like the "murderer's bloody fingerprint" which asserts the reality that makes art irrelevant for the moment. Seen in this context, the ornamentation and the mirrors point to Herman, Lydia, and Ardalion's self-absorption, when Nazism is on the rise. Mirrors were a favorite device of Fassbinder since they suggest performance and self-absorption. In another scene, Hermann and his double make arrangements inside a glass house while a storm is brewing outside. I interpret the storm as a metaphor for the political storm building up.

Despair is typical of Fassbinder in that there is no reason for Nabokov's novel to have been adapted to make this kind of statement about the self-absorption of the bourgeoisie at the time of the rise of National Socialism, and Stoppard is therefore right in his assessment that Fassbinder does not care about Nabokov's novel at all. But one gets the sense from his films by this time that his purpose not being understood even by his collaborators was something he had become accustomed to; he was getting by, as if in perpetual disguise, never revealing himself at all. Here is his actor Peter Chatel on an occurrence pertaining to the press conference around *Despair*:

> Before going to the press conference, I asked Rainer some questions that were likely to come up, like there were so many Nazis in the beginning and none later on. Rainer said "I forgot to shoot that." Was that what I was supposed to say? "No," said Rainer, "you tell them that the Nazis are in the beginning because in 1933 people were much more aware of them, but they got used to their presence and they weren't noticed any more."[32]

As indicated, while dealing with the history of German cinema, Germany went through a tumultuous period because of the War and its aftermath, and Fassbinder grew up in those years. Many of Fassbinder's attitudes can be laid at the doorstep of that period, like his left-wing beliefs, but with a lack of political faith in any kind of utopia, in fact an active distrust of political programs. Ultimately, Fassbinder's politics can perhaps be subsumed under the rubric of "left-wing melancholy" although the term is imprecise. It refers broadly to a way of life that recognizes the exploitation in society but is unable to envisage the possibility of fundamental changes. It is a deeply felt anguish but in Fassbinder's case it showed up as covert mockery and irony.

NOTES

1. "It would be naïve to accept everything Rainer Werner Fassbinder told interviewers about his childhood." Ronald Hayman, *Fassbinder: Film Maker* (London: Weidenfeld and Nicolson, 1984), p. 1.

2. In 1910, distressed church observers of German cinema had counted 97 murders, 51 adulteries, nineteen seductions, 35 drunks, and 25 prostitutes in as few as 250 films. Hayman, p. 16.
3. Heide Fehrenbach, *Cinema in Democratizing Germany* (Chapel Hill: The University of North Carolina Press, 1995), p. 53
4. Fehrenbach, p. 55.
5. Fehrenbach, p. 59.
6. Ronald Hayman, *Fassbinder: Film Maker*, pp. 1–13.
7. Thomas Elsaesser, *Fassbinder's Germany: History, Identity, Subject* (Amsterdam: Amsterdam University Press, 1996), p. 45.
8. I have Harold Pinter's plays like *The Dumb Waiter* (1957) and *The Birthday Party* (1957) in mind. A film thriller that works on this same notion is Jean-Pierre Melville's *Le Samourai* (1967).
9. Aristocracy and landed gentry (*Effi Briest*), haute-bourgeoisie (*The Bitter Tears of Petra von Kant*, *Martha*), old money (*Chinese Roulette*) and nouveaux riche (*Lola*), show-business circles (*Lili Marleen*, *Veronica Voss*), petty bourgeoisie (*The Merchant of the Four Seasons*), working class (*Mother Kuster's Trip to Heaven*), lumpenproletariat (*Love Is Colder than Death*), "Gastarbeiter" (*Katzelmacher*), Black people—both foreign and indigenous (*Fear Eats the Soul*, *Pioneers in Ingolstadt*). Also as wide are the professions: journalists, industrialists, property tycoons, shift workers, intellectuals, writers, office workers, trade unionists, fanners, shopkeepers, butchers, bartenders, pimps, prostitutes of either sex, petty gangsters, contract killers, dealers, sailors, soldiers, and soldiers of fortune. Thomas Elsaesser, *Fassbinder's Germany: History, Identity, Subject* (Amsterdam: Amsterdam University Press, 1996), p. 19.
10. Elsaesser, p. 21.
11. The praise that an actor is "unrecognizable" in a role (for instance Gunnar Bjornstrand in *The Seventh Seal*), used with regard to a Bergman film when Bjornstrand plays other kinds of roles as in *Wild Strawberries*, would not generally be applicable to a film by Fassbinder.
12. I have in mind John Berger's maxim that nakedness is the abandoning of disguise while nudity is having one's very skin and hair being turned into a disguise. John Berger, *Ways of Seeing* (Harmondsworth: Penguin, 1972), p. 54.
13. Characteristics identified by Sergei Balukhatyi (*Poetics of Melodrama*) and quoted in Daniel Gerould, "Russian Formalist Theories of Melodrama," in *Imitations of Life: A Reader on Film, Television and Melodrama*, ed. by Marcia Landy (Detroit: Wayne University Press, 1991), pp. 121–9.
14. Peter Brooks, *The Melodramatic Imagination: Balzac, Henry James, Melodrama, and the Mode of Excess* (New York: Columbia University Press, 1985), p. 15.
15. Thomas Elsaesser, *Fassbinder's Germany: History, Identity, Subject*, pp. 33–34.
16. "The viewer should be able to activate things and feelings in himself via the characters, but the structure of the presentation ought to give him the possibility for reflection [...] that is, the mise-en-scene ought to be such that it makes distance and reflection possible." Fassbinder quoted in Thomas Elsaesser, *Fassbinder's Germany: History, Identity, Subject*, p. 34.

17. "Ali and Emmi suffer from ostracism because of a liaison that is considered a breach of decorum. But the way it presents itself to the couple is as a contradiction: they cannot be seen together, because there is no social space [...] in which they are not objects of aggressive hostile disapproving gazes [...]. Yet conversely they discover that they cannot exist without being seen by others, for when they are alone, their own mutually sustaining gaze proves to be insufficient to confer on them or bestow them a sense of identity." Thomas Elsaesser, *Fassbinder's Germany: History, Identity, Subject*, pp. 65–66.
18. There is self-consciousness in Fassbinder's stylization when he often quotes cinema as noted by Elsaesser: "As a modernist form of deconstruction, this draws attention to the artifice of representation, or critically comments on the claustrophobic, stifling and cluttered world in which these characters are at home. One remembers Sirk's frequent mirror shots in the opulent mansion of WRITTEN ON THE WIND and the final tableau with a picture-window in Rock Hudson's cabin in ALL THAT HEAVEN ALLOWS. But Fassbinder also cites the cinema itself, whether by 'fetishizing' its apparatus and thus barring us from the pleasure of unmediated transparency to which the director himself is nonetheless attached." Elsaesser, pp. 58–59.
19. "Ali and Emmi's romance is prescribed to have no hope, to be finite, whether or not the social world they inhabit reacts with explicit and overt racism, as it does in the first half of the film, or with disingenuous approbation, as it does in the second half of the film. The ending of the film sees Ali and Emmi reunite, albeit through the terms of corporeal failure and physical pain." The above description is very different from the one I have offered, that Emmi is herself implicated in the racism of their social world. Elena Gorfinkel, "Impossible, Impolitic: Ali: Fear Eats the Soul and Fassbinder's Asynchronous Bodies," in *A Companion to Rainer Werner Fassbinder*, ed. by Brigitte Peucker (Chichester: Wiley-Blackwell, 2012), p. 503.
20. "Equally ingratiating is the Gastarbeiter's own social Darwinism: only too ready to complain when a Turk is hired by his firm, Jorgos turns out to be as much a racist as his German tormentors. This final irony neither explains nor excuses anything, and yet it closes the film on a suitably sour note." Thomas Elsaesser, *Fassbinder's Germany: History, Identity, Subject*, p. 46.
21. "The film was clearly received as an allegory of West Germany. *Maria Braun* no doubt also helped fill a gap that, at least for American audiences, had been there almost since the 1930s: cinematically speaking, the world knew nothing about Germany, nor how Germans felt about fascism, defeat and reconstruction. Thus, the most commonly accepted reading of Fassbinder's film saw Maria Braun as standing for the pragmatic, post-Hitler Germany. This treatment of a woman who picks herself up from the ashes of a war and national disgrace and becomes an efficient, modern business woman is a dramatic metaphor for a nation making a similar transition. The divided nation's astonishing economic success was, however, bought by a double denial. Predicated on disavowing the fascist legacy and thus periodically visited by a 'return of the repressed', the new Germany seemed seized by

nameless bouts of melancholy, indicative of deeper ills, despite material prosperity and outward signs of stability. Having denied itself the fulfilment of emotional needs, sacrificing spiritual values for status and wealth, it wittingly or unwittingly colluded with Cold War rearmament, in which former enemies became allies, and pacts were made, along with fortunes." Thomas Elsaesser, *Fassbinder's Germany: History, Identity, Subject*, p. 102

22. Elsaesser, p. 97.
23. Robert Katz, *Love Is Colder than Death: The Life and Loves of Rainer Werner Fassbinder* (New York: Random House, 1987), p. 93.
24. Thomas Elsaesser, *Fassbinder's Germany: History, Identity, Subject*, p. 23.
25. Ronald Hayman, *Fassbinder: Film Maker*, p. 116.
26. Norbert Jiirgen Schneider, *Musikdramaturgie im Neuen Deutschen Film* (Munich: Olschliiger, 1986), p. 187, quoted in Thomas Elsaesser, *Fassbinder's Germany: History, Identity, Subject*, p. 108.
27. Paul Thomas, "Fassbinder: The Poetry of the Inarticulate," *Film Quarterly*, 30.2 (Winter 1976–77), 4.
28. Brigitte Peucker (ed.), "In *Despair*: Performance, Citation, Identity," in *A Companion to Rainer Werner Fassbinder*, p. 291.
29. Peucker, p. 291.
30. Peucker, p. 301.
31. Walter Benjamin, "The Author as Producer," in *Understanding Brecht*, trans. by Anna Bostock (London: Verso, 1998), p. 95.
32. Peter Chatel as quoted in Robert Katz, *Love Is Colder than Death: The Life and Loves of Rainer Werner Fassbinder* (New York: Random House, 1987), p. 118.

4

Robert Altman: History, Ideology, and Generic Revision

Robert Altman can be a difficult filmmaker to write about because of his sheer prolificity and variety. Buñuel and Fassbinder were also enormously productive but it is easier finding unity in their work that in Altman's, whose work at first glance also seems more straightforward. Altman came into his own in the 1970s, perhaps the golden age for Hollywood. Breaking through at this time of upheaval, Altman used the newfound freedom of the Hollywood filmmaking climate to establish himself as the critical commentator and alternative historian of the United States.[1] This latter assertion partly draws on the recognition that historical memory is retained in cinema through genres such as Hollywood films that make myths of historical moments.[2] A well-acknowledged example would be the western, which mythologizes the origins of the American nation-state. Altman set about systematically subverting these genres and his generic revisions can therefore be seen as a kind of alternative history-writing.

Altman worked for a long time before he became renowned. He directed some 65 industrial films and documentaries before being hired by a local businessman in 1956 to write and direct a feature film in Kansas City on juvenile delinquency, and this suggests someone who had become accustomed to taking up routine projects before he became an auteur in cinema. There is therefore a vast gap between his best work and many of his other films, which can be indifferent. It also means that he associated with the establishment although, given his independence, the alliances were uneasy.[3] Altman worked best in the 1970s when the climate was right for him[4] and most of the films examined are from that period. Given that his favorite vehicle was generic revision, this essay rides on the sense that he was largely debunking dominant American myths, which would amount to satire or parody. Still, the film that brought him fame as a satirist *M*A*S*H* (1969) is satire but more irreverent comedy and I will attempt to explain why. I divide this enquiry into Altman's cinema into two sections beginning with his generic subversions and then his arthouse films; but before doing that, I will examine *M*A*S*H* briefly since it established his distinct voice. Needless to add, only a small minority of his films can be studied here.

*M*A*S*H* (1969)

This film is set in 1951 and the Korean War and the title refers to the 4077th Mobile Army Surgical Hospital and the adventures of three surgeons, Capt. Benjamin Franklin "Hawkeye" Pierce Jr (Donald Sutherland), Capt. Augustus Bedford "Duke" Forrest (Tom Skerrit), and Capt. John Francis Xavier "Trapper John" McIntyre (Eliott Gould). The three doctors have little respect for military protocol, having been drafted into the Army, and are prone to mischief, chasing after women, and heavy drinking but they are excellent surgeons. Being irreverent they come into conflict with Major Frank Burns (Robert Duvall) and the newly arrived Major Margaret Houlihan (Sally Kellerman), who is nicknamed "Hotlips." The funny bits are Burns and "Hotlips," both of who are devout Christians, "making out" and their lovemaking recorded and played out aloud in the camp. In another segment, "Hotlips" is taking a shower and the sides of the bath tent are lifted up to expose her to the public. The film has a great deal of "violence" in the form of blood seen on the operating table, but we never see the war and nothing about the enemy is even hinted at. The high point of the film is a football game with 325th EVAC Hospital with General Hammond in charge. By this time the conflict with "Hotlips" is forgotten and she is also busy cheering their side, although for the wrong reasons.

Essentially, *M*A*S*H* is celebrating irreverence toward authority and all three surgeons are alike in this way. But since the film is dealing with the grave issue of saving human lives, it is portraying its protagonists as highly competent, and their competence and confidence as surgeons justify *their* irreverence. Surgeons are badly needed in war and this means that their frequent "arrests" for indiscipline or insubordination need not be taken seriously. There are no crises, the three protagonists breezing through situation after situation, their playfulness not tested. The film was a big success and one attributes it to its impudent stance, which appealed to the young, and if the film had even been two years later or earlier it might have bombed.[5] It is nominally an anti-war film but—unlike Kubrick's *Full Metal Jacket*—it has little to say either about the actualities of war. To recall a quote from Linda Hutcheon used in the chapter on Kubrick, it notes that the ironic effect achieved is inversely proportional to the number of overt signals used to achieve that effect and I think there are too many overt signals—like Hotlips being openly mocked by others—that undermines the irony and satire. At the same time, if we put aside the irreverent protagonists of the film and their actions, the film is satirizing many aspects of the military (like sporting rivalries, religious platitudes) and it is only the fact that Altman has these irreverent characters *who are doing the mocking in the film* alongside himself (as the director) that could confuse us.; I described the film more as comedy than satire since the irreverence is less an attribute of the authorial vision than of the portrayed characters. *M*A*S*H* was a hit with young audiences because they could identify with irreverence—especially when personified in the protagonists—but the irony is a different matter and depends on the interpretation to which the film must submit. Most of his other films were commercially unsuccessful, except *The Player* (1992) which satirizes Hollywood's ways and uses stars to appear playing themselves without offending anyone (as Altman himself declared).

An aspect of *M*A*S*H*, for which Altman became renowned, needs comment here and that is his use of overlapping dialogue. It is difficult to keep track of what is being said since so much is spoken together, but when studied it at length we realize that we are not required to. The strategy goes along with the choreography, people preoccupied with their own trajectories unlike in film drama where everyone is focused on teleology and participates in the furtherance of a single story. In *M*A*S*H*, it is essential to downplay the war drama, this amounting to cheekiness that parallels that of the three key protagonists. To elaborate, it would have been unthinkable to stage "irreverence" dramatically and the overlapping dialogue helps.

In Altman's generic subversions dealt with in the subsequent section, the choreography, used to suggest randomness, as well as the overlapping dialogue, becomes a way of interrogating historical "drama," which has been the chosen method of genre cinema, as for instance, in the larger-than-life acts of John Wayne. If multiple, (often) undramatic events are happening together in an Altman film, it is negating the usual choices of genre cinema where there is a selective dramatization of events. This has a parallel in the formal writing of history where much is cast aside to further teleology. The revisionist agenda in genre cinema is to question the dominant mythmaking; but, where telling a dramatic generic tale with a contrary discourse (Arthur Penn's anti-western *Little Big Man*, 1970) would only make a counter-claim upon history, Altman's strategy interrogates the construction of history itself. Altman's deliberately incomplete resolutions and open endings also work against the sense of a definite teleology. If Buñuel subverts teleology to not play God, Altman subverts it to not play historian.

Generic subversions

Altman worked in a multiplicity of genres and I will examine only four films; his anti-western *McCabe and Mrs Miller* (1971), the inversion of the hard-boiled detective story in *The Long Goodbye* (1973), his take on the noir/caper film in *Thieves Like Us* (1974), which is a remake of Nicholas Ray's *They Live by Night* (1948) and from the same sub-genre category as *Bonnie and Clyde* (1967), and his comic strip hero film *Popeye* (1980), which anticipated later exercises like Steven Spielberg's *The Adventures of Tintin: The Secret of the Unicorn* (2011).

McCabe and Mrs Miller (1971)

The film is set in wintery settings, a mining settlement in the north-west named Presbyterian Church into which enters a stranger on horseback. Before entering the settlement, he dismounts, takes off his furs, revealing himself to be dressed in a suit and he puts on a bowler hat. In the local saloon owned by Sheehan, he sets up a card table and, since he is conspicuous, word gets around that this is John "Pudgy" McCabe, who shot Bill Roundtree. No one knows who Bill Roundtree was, and a bystander who claims to know says he was a Governor somewhere. McCabe (Warren Beatty), who is accosted with the information neither confirms nor denies it but in response to the question of whether he is a gunfighter, describes himself as a "businessman."

McCabe seems to have the makings of a shrewd businessman and sets up a makeshift brothel purchasing three prostitutes from a pimp in the nearby town of Bearpaw for $200 and the three operate from tents with placards displayed outside announcing their respective prices. With his reputation preceding him, McCabe becomes a dominant citizen in the settlement and a British Cockney woman Constance Miller (Julie Christie) arrives with a string of girls with a business proposition for him: starting a better-class brothel with clean sheets, with the two of them sharing the income, and a saloon of McCabe's own alongside. Constance Miller is a "whore herself" but better educated, can keep accounts, and knows how to handle the girls and keep them perpetually busy: "A whore who has time to think has her thoughts going to religion."

Altman's gambit in the film is evidently to set up a vision of the west different from that traditionally offered. Where the protagonists are usually gunfighters or cowboys, McCabe is a small businessman and where the setting in a western is usually a dry, sunny area with wide open spaces, Presbyterian Church is cold and wet; it is raining or snowing and people walk about huddled in their jackets. The overlapping conversation is amusing, although people are saying unimportant things, like "There are Chinese whores in Bearpaw and one of them is an actual princess. Amos Lindville paid $5 to find out and a guy like that wouldn't spend $5 on nothing."

As the settlement becomes more prosperous so does the brothel, and Mrs Miller stocks it with more girls, some Chinese. A local miner Bart has ordered a "mail order" bride and Ida (Shelley Duvall) arrives for him. When he dies, she becomes a whore also—she was doing her duty with Bart and this will be a duty as well. The prostitutes double in other kinds of work and when one refuses a client because she's a cook and not a whore, the client exclaims that that is a waste of good experience. With the Chinese workers also coming, Mrs Miller locks herself in with opium.

Nearly half the film goes by before two agents Sears and Hollander arrive at Presbyterian Church offering to buy McCabe out for their client the Harrison Shaughnessy Mining Company with an office in Bearpaw and offer him $5500 for all his holdings. They have bought the zinc mines and also Sheehan out for $1500. McCabe behaves the way he thinks an active businessman should, offering the two drinks and being hospitable but giving their words scant attention. He puts on the air of someone too preoccupied to listen to them. He is dealing with a different kind of clientele but he seems oblivious. Mrs Miller is aghast when he tells her and says Harrison Shaughnessy will "sooner put a bullet in your back than bargain and negotiate." Later, Sears and Hollander come back with a better offer of $6250 but McCabe tries to set them up with whores and invites them to breakfast the next morning, quoting a tentative $15,000 for his holdings. McCabe does not understand the way a monopoly business functions and expects the two agents to be open for more bargaining but they do not return. Sears was inclined to go back but Hollander, who is furious, called McCabe a fool and was opposed. These scenes are innocuous at first glance but become chilling when we realize that this professional discussion among the agents concludes with the passing of a death sentence on McCabe.

McCabe and Mrs Miller although quiet is mordant as irony and not amiably irreverent as *M*A*S*H* was; the latter half is sinister, conveying the sense of McCabe unable to do much to sidestep his fate. When Sears and Hollander are not reachable, a lawyer he meets lectures him about the importance of the small businessman to America. The chief irony in the film evidently owes to this rhetoric about the small businessman having nothing to do with the actual situation faced by McCabe, although it is not comic but terrifying. A harmless cowboy (Keith Carradine) wanting to use the brothel is briefly mistaken for a killer and everyone is tense about what is going to happen. When the killers do appear, there are three of them and McCabe wants to make a deal. Unfortunately, says Butler, their leader, they cannot make deals and are only sent in when a deal cannot be made. The locals are spineless and Sheehan has told Butler about "Bill Roundtree." McCabe has been riding on his reputation as a gunfighter and Butler humiliates him in Sheehan's saloon, "Bill Roundtree was my best friend's best friend and you killed him." The closing of the film is the familiar shootout from the standard western (like *High Noon*, 1952) but with an entirely different significance, and with no possibility of the right triumphing, since McCabe is dealing with a corporate and not individuals.

McCabe is not the western hero taking a stand but a fraud with an undeserved reputation that he used to get himself a good position. He bluffs his way through the film only to find that in the end his pretenses cannot pay off. By revealing the human weakness of his character, Altman destroys the mythical image of the westerner[6] but also changes its context. The primary subversion here is that "America" was not created when the white man fought Native Americans or when the law prevailed over lawlessness but when monopoly houses consumed the small entrepreneur.

These are aspects of the westerner[7] that he subverts: where he is a man of leisure McCabe is always busy; whereas the westerner stands alone, McCabe is a man trying to get on in the community through his generosity with liquor and homespun homilies that no one comprehends ("If a frog had wings, he wouldn't be bumping his ass so much.")[8] but that pass for the superior wisdom of an experienced man. The woman he loves is not the embodiment of Christian virtues but a prostitute who charges him. Lastly, when McCabe dies as heroically as he is capable of, there is no one to watch and Mrs Miller is in an opium-induced dream. The death of the western hero, we recollect, was often commemorated in the history of the nation as in John Ford's *The Man Who Shot Liberty Valance* (1962) and McCabe will remain unsung. This is satirized when the lawyer with political ambitions hails McCabe as the likely national hero of small business and will make headlines in the *Washington Post*. The cowardly folk in Presbyterian Church—Sheehan and the priest who refuses McCabe shelter—offer no resistance and this is a darker rendering of a comparable situation in *High Noon* where support was still to be had

The Long Goodbye (1973)

The Long Goodbye—based on a novel of the same name by Raymond Chandler—belongs to a genre called the "hard-boiled detective story" which may be regarded as a sub-category under noir and includes films like John Huston's *The Maltese Falcon* (1941) and Roman Polanski's *Chinatown* (1974). This kind of detective story, which developed in the 1920s through short action stories in pulp magazines like *Black Mask*,[9] differs from the whodunit in that the puzzle being disentangled is dynamic. The detective solves the mystery not through the powers of his intellect but through encounters and collisions, transforming the contours of the conspiracy in the process of unraveling it. The crimes are not premeditated acts of evil but responses to immediate social stimuli. This means that when the "story" finally emerges through its reconstruction in the reader's consciousness, its significance extends much further than one anticipates in a detective story and contains a political vision usually of a pessimistic kind.

Recounting the narratives of Altman's films is not easy because of his deliberate attempts at lowering the drama and obscuring the plot. As already indicated, there is movement and dialogue that amount to an excess since Altman is questioning the definite teleology that genre films opt for and the satisfaction of a neatly rounded ending is usually denied. In the first part of *The Long Goodbye*, the private eye Philip Marlowe (Elliot Gould) is woken up by an old friend, Terry Lennox (Jim Bouton), who has just had a violent argument with his wife—his face is scratched—with a request that Marlowe drive him to the Mexican border. Marlowe obliges but is arrested the next morning because Sylvia Lennox has been beaten to death and Marlowe has notionally abetted in a murder. Marlowe is released three days later because Terry Lennox killed himself in Mexico, leaving behind a full confession. A day later, a rich woman named Eileen Wade (Nina van Pallandt) hires Marlowe to find her husband, an alcoholic writer by the name of Roger Wade (Sterling Hayden), who has been missing for several days. Marlowe's refusal to betray his friend Lennox is now public knowledge, and this probably convinces Eileen Wade that Marlowe can be trusted. Marlowe traces Roger Wade to a private clinic run for drug addicts and alcoholics by a crooked doctor named Dr Verringer (Henry Gibson). Marlowe escorts Roger Wade home leaving an unpaid bill of $5500 still to be settled with the doctor.

The character of the hard-boiled detective is crucial to the genre and the protagonist in the story is usually ahead of the police. In Altman's film, Marlowe lags behind the police in most ways although he hits upon the real solution eventually. This is not to say that Marlowe is a comical bungler but, rather, that the private eye operating alone is something of an anachronism in contemporary LA. Moreover, the social structure hinted at by Altman is so complex and muddled (rather than corrupt as Chandler might have had it) that no entity, not even the police, has a full measure of it. Instead of relying on motivation, therefore, Altman sets up three or four scenes that hit the spectator hard largely because they defeat explanations.

There are several threads for Marlowe to unravel: it turns out that the Wades knew the Lennoxes socially since they lived within the same gated community. Another is the gangster Marty Augustine who lost $355,000 that Terry Lennox was carrying for him into Mexico but did not arrive. Augustine is not a gangster in the Corleone mold but a psychopath who has henchmen doing his every bidding. The first occasion we see him, he comes with his hoods and his young girlfriend Jo Ann Eggenweiler (whom he "loves") to talk to Marlowe about the money, and she asks for a Coke. When a nearly empty bottle is brought, Marty finishes it himself and smashes the girl across the face with the empty bottle, following it up instructively to Marlowe: "I love her and can do this. You, I don't even like." The episode is shocking in its suddenness and, later, we hear Marty Augustine confessing to how deeply he regretted hurting the girl he loved (He "sleeps with other women" but "makes love only to her"); but he stripped naked in the hospital as a token of his honesty and apologized and the girl confirms the story, optimistic through her bandages. In a party scene at the Wades later, the drunken writer is being boisterous when the diminutive Dr Verringer turns up and demands the money that Wade owes him; Wade (gigantic beside Verringer) initially laughs it off derisively but a ringing slap across his face from the dwarf and he succumbs, muttering. We do not know the hold the small man has on the writer but, even as the guests are looking on, Wade has a collapse and weakly writes out the cheque.

It is perhaps this disinclination to ascribe clear-cut motives to conduct that takes *The Long Goodbye* away from what we understand as American noir. Where everything is eventually cleared up in the hard-boiled detective story regardless of the twists, Altman insists that many things are not knowable. The complete resolutions in the genre testify to a social structure of causes and effects that is cogent but Altman subverts that. While being deeply sardonic about social behavior, the film does not make such inexplicable behavior seem incredible and we are left to wonder. Why some people exert such power over others is an issue but the opaqueness of power structures justifies Altman's approach. The relationship between Marty Augustine and Jo Ann, for instance, is grotesque but it nonetheless seems stable and hardly incredible.

The traditional hard-boiled detective story, for all its pessimism, deals with a socio-political world populated by rational individuals, and evil owes to their machinations (as in *Chinatown*), but *The Long Goodbye* blocks that systematically, which is why even a satisfying denouement tying loose ends together is denied. It finally seems that Marlowe is looking for a *personal solution* for himself rather than seeking out satisfying explanations. Phillip Marlowe in Chandler's stories is untainted in a corrupt world but Marlowe here is closer to McCabe from the earlier film, in that he is only working for himself and not righting a system gone bad. The rational structure of a Chandler novel (like *The Long Goodbye*), for all its pessimism, shows faith that the system is rationally devised and that is something that Altman's ironic take does not allow.

Thieves Like Us (1974)

Robert Altman is a "realistic" filmmaker and, in this way, differs from mainstream Hollywood. Hollywood has generally tried to create "drama" around individuals, or rather, "individualities" as types. The factor bestowing a type with "individuality" is her/his separateness from other people, her/his ability to make us believe that she/he is as "we" are behind our disguises, someone capable of "defeating our self-defeats."[10] There is close identification with the star-as-protagonist because she/he represents us as we might have been if we had had the strength to be what we actually are. We, therefore, project ourselves into the "individuality as type," something we do not do with "character types." In *Spider-Man* (2001), for instance, Peter Parker (Tobey Macguire) is the "individuality," while his uncle and aunt are character types. The "individuality" is valued because she/he stands out above her/his given social role to which the character-type submits. One way of achieving identification with him/her is to catch the action from the subjective viewpoint of the individuality but not from that of the character type.

Most Hollywood filmmakers including legends like Scorsese adopt this strategy of demarcating the individuality from the character type and that is where stars like Robert De Niro fit in. Kirk Douglas as a towering presence in Kubrick's *Paths of Glory* owes to this deliberate demarcation. American genre films use stars and a major difference between Altman's genre films and those he is "subverting" is that Altman either does not use stars or he subdues their presence as he did with Warren Beatty in *McCabe and Mrs Miller* or Elliott Gould in *The Long Goodbye*, where the star is not privileged. Elliott Gould in the latter film is very different from Humphrey Bogart as Phillip Marlowe in Howard Hawks's *The Big Sleep* (1946). Altman's use of random movement within spaces and overlapping dialogue[11] also works against the Hollywood approach since it does not privilege the star over the others.

Thieves Like Us is a remake of Nicholas Ray's *They Live by Night* and (mainly) tells the love story of Bowie (Keith Carradine) and Keechie (Shelley Duvall) which begins in 1936 just after the depression era. Bowie, a young man of 23 convicted of murder when a teenager, and Chicamaw, an old offender, escape from a Mississippi prison in 1936 and join up with another convict, T-Dub. They stay with an acquaintance that owns a garage and set to work robbing banks. Bowie gets to talk to Keechie, the garage owner's daughter. They next hole up with T-Dub's sister-in-law Mattie and her children, including her older daughter Lula, the object of T-Dub's lascivious attention. Mattie's husband is in jail and Mattie is hoping to get him out on parole as she manages the house with her guests and her three children. The film does not use music but instead relies on actual radio programs of the era used appropriately, including a dramatic one about a vigilante named "the Shadow" who is fighting the underworld, a radio version of Romeo and Juliet, and political speeches (including those of Franklin Roosevelt) pertinent to the period. The film belongs to the same genre as Arthur Penn's *Bonnie and Clyde* (1967) but their differences are instructive. That film used stars in the key roles and the characters live and die heroically. *Bonnie and Clyde* is a counter-culture film, and robbing banks is presented as a radical activity for an era when banks were especially seen to be foreclosing and seizing mortgaged property. In Altman's film, the lead actors are ordinary and blend into the scene, attracting no more attention than the little-known actors who play T-Dub and Chicamaw.

The archetypal plot, which is used to different ends in the sub-genre to which *Thieves Like Us* belongs, has to do with a young man striving to rise above the milieu through crime and coming into conflict with the law. The romance is a consequence of the man attracting the attention of the woman by not being ordinary. In Joseph H Lewis's *Gun Crazy* (1950)—which is from the same genre as Altman's film—Bart is an excellent marksman who defeats carnival sharpshooter Annie and she is smitten with him. In *Bonnie and Clyde*, Bonnie Parker sees Clyde Barrow rob a store and falls in love with him; in *Badlands*, Kit resembles James Dean and tries to make the most of it through his stylish manner and Holly falls in love with him. There is a sense of people trying to escape their "ordinariness" and their violence becomes a statement in that direction. The tension the films create is therefore between the quest for individuality and its suppression by the state, although it is only in *Bonnie and Clyde* is the state equated with social oppression.[12] The turning point in each story is an excess committed by the protagonist, usually a wanton killing, and the ending is tragic in most films, with both protagonists killed. *Badlands* opts for an anthropological take on the theme by showing the two as products of popular culture and hence discards the tragic finale of the other films.

In Altman's film, there is so much going on—like the family circumstances of those who shelter the three—that this discourse of individuality in conflict with the state is dropped since the state is not a monolith (as it is in *Bonnie and Clyde*). In a tragicomic episode that illustrates this, Bowie drives into prison pretending to be a lawman to get Chicamaw out. The captain, whose permission he needs, is having his lunch and Bowie waits in the car outside; when the captain comes out, he is unwittingly led to help Chicamaw and Bowie escape, with himself as hostage. But what really upsets the captain is that he is retiring and this black mark will affect his pension. Chicamaw is supposed to tie him up but he kills the captain instead. After the episode, Chicamaw, rather than being grateful, is cross that a country boy like Bowie should pull off something like this, and the annoyed Bowie abandons him. The love story of Bowie and Keechie must be among the most convincing and tender in all American cinema and the events leading to Bowie's killing are matter-of-fact although its finale is gruesome. The law catching up with him is not relentless but Bowie is too much of an irritant and there is no other way out. Bowie, with every success, has got bolder and bolder and he is bound to exceed himself. Keechie, who is expecting a baby, is later shown to have gotten over Bowie's death.

Using the word "irony" to describe *Thieves Like Us* may seem misplaced since it is so warm in attitude toward its characters. But when we see it in relation to the noir category it is subverting, the epithet begins to seem reasonable. Noir is a dark category and a version of the same tale was also told by Fritz Lang as *You Only Live Once* (1937) using expressionist imagery; the underlying theme is, once again, individual aspiration coming into conflict with the state, doomed love as an attendant motif. The sub-genre is dramatic; and sharply foregrounded is a tragic story in which the conflict is mourned. Altman, in contrast, gives a warmly realistic (though violent) picture of the same story that is only one among many within the film. The story of Lula's family and T-Dub—which intersects that of Bowie and Keechie—could be another and even the captain's is a story on its own. The term "subversion" used with regard to Altman's genre reworking implies a harsh approach but with genres that are inherently dark and obsessive as noir Altman subverts them differently. But it is still subverting a dominant mythology around individuality and the state that is peculiarly American.[13] If one were not tuned to other films of the genre (like Nicholas Ray's), the film would not be "ironic" but its irony lies more in the expectations created by the genre's teleology being denied. I tried to show that Luis Buñuel's realistic films often produced irony through the same strategy.

Popeye (1980)

This film was Altman's attempt to reproduce the world of the animated cartoon using real actors and without using tricks like speeded-up motion. To understand the world of the animated cartoon, it can be usefully compared to silent film comedy, that of Harold Lloyd, Charles Chaplin, and Buster Keaton: the camera confirms the reality of what is happening, but the special properties of silent film—silence itself, and a speeding up of action—transform that fact into fantasy, a "fantasy of fact"[14] in the sense that the real is given a slightly unrealistic air that supports the kind of action represented by the acrobatics that Keaton performs in *Sherlock Junior* (1924). With animated films, it was the reverse that actually happened:

> In silent cartoons, fantasy in the form of blatant impossibilities was the stock in trade, so that, for example, Felix the Cat could detach his tail and use it as anything from a spyglass to a grappling hook. What was missing was a foundation in fact that would permit audiences, with a single suspension of disbelief, to accept the reality of what was happening in the cartoon they were watching. Even the most distinctive characters in silent cartoons were like playing cards in a magician's trick; they made no claim on the imagination outside the bits of cleverness in which they were used.[15]

What Altman tried to do in *Popeye* was to provide a "foundation in fact," create that coherent world in its entirety and not simply reproduce a magician's tricks.[16] He had a set "Sweethaven" constructed in Malta and populated it with characters taken from the cartoons: Popeye (Robin Williams), Olive Oyl (Shelley Duvall), and her family, Swee'Pea, Poopdeck Pappy (Ray Walston), Bluto (Paul L Smith), Wimpy (Paul Dooley), Ham Gravy (Bill Irwin), and added several more. The film is a musical with songs by Harry Nilsson. Breaking up the narrative with songs helps the absurd air cultivated in the film and justifies the episodic quality that cartoons have. As already indicated, the overlapping dialogue and the random detail Altman brings into his films make them episodic, but *Popeye* not being a realistic film, creating the sense of catching random action was not an option. The songs are used instead to punctuate the film and help undermine causality. Sweethaven is a self-contained world with a "national anthem" modeled on "Star Spangled Banner," the first song sung by its citizens when they go about their morning chores. Doing *Popeye* like this is more difficult than doing *Tintin*, which is a quasi-realistic adventure, where the main achievement is to doctor the images to resemble Hergé's drawings.

In *Popeye*, a stranger comes to Sweethaven, rowing across the sea, to be greeted by the anthem and accosted by a tax collector who charges him "docking tax," "new-in-town tax," "rowboat-under-the-wharf tax," "leave-your-junk-lying-under-the-wharf tax," in all amounting to $1.87. He is collecting on behalf of the "Commodore" and when the stranger wants to know who the "Commodore" is, the tax collector threatens to charge him with a further "question tax." The stranger pays in a variety of currencies apart from the dollar: a peso, a guilder, a red cent, and a franc among others; the stranger is apparently a sailor who has been around. A stranger in town is a frightening new development and people close their doors and windows to him until he finds lodgings with the Oyl's, whose daughter Olive is getting engaged that evening to Captain Bluto. The stranger is Popeye and, en route, he buys a bunch of "carricks" (carrots) on the way and avoids spinach. The vendor, who has a thick European accent, and appears Jewish, quotes $1.50 (revised from a dollar), Popeye pays him a nickel ("I pays what I feels like paying"). In the ramshackle room provided by the Oyls, escorted to it by their daughter Olive, he takes his belongings out of a bag and kisses what we take to be a picture in a frame. When we are shown what is in the frame, we see it is empty but has "Me Poppa" crawled inside; Popeye (who is an "orfink") has never known his father and has sensed that he is in Sweethaven.

Popeye works brilliantly in parts but tends to fall flat in sections largely because the comic strip rhythm is hard to keep up. The opening segment tries to suggest a busy but zany space: a man chasing after his hat lying on the road but kicking it further and further, Wimpy setting out chewing on his burger, people closing their doors and windows on the stranger in town, the Oyl's residence with "Rooms for Rent" with the "S" crossed out revealing that they have only one room available. Olive is due to be engaged to Captain Bluto that evening and sings an optimistic song but keeps coming back to "he's large," his only positive attribute. Captain Bluto is the informal authority in town and he is collecting taxes on behalf of the "Commodore." The Oyl family is favored and Popeye is granted exemption from further taxation. Cole Oyl is the father in the Oyl household, Nana Oyl the mother, and Castor, Olive's younger brother.

Popeye is excluded from the engagement party that evening since he does not have an invitation but Olive is also running away. Bluto arrives late and waits for Olive while the other guests fawn upon him; but in the street below, Popeye and Olive come upon a basket with rattling coming from within. When Popeye, who suspects a rattlesnake, opens it, they find a baby with a rattle. Its mother has left a note pleading "financial obligations" and promising to come back in 25 years to reclaim the child; comic strip characters, after all, do not grow older. Popeye becomes its "mudder" and names the child "Swee'Pea." When Bluto comes upon Popeye, Olive, and Swee'Pea, he literally "sees red" (frame with all three dressed in red, the wall and curtains also in the same tint). From that day on, the Oyl family loses all privileges and they are ruined. Bluto also destroys their house and beats up Popeye. Popeye spins so swiftly on being hit that he drills a hole through the wooden pier and falls into the sea.

Altman's film succeeds in building a milieu with the logic of the comic strip with its own underworld and law enforcement, both run by the Commodore through Bluto. Swee'Pea has the ability to predict the future (he whistles at the right choices) which means that he can help Wimpy at the racetrack (half a dozen mechanical horses with tawdry women hanging about). The sternly moral Popeye takes exception to bring the child to a "place of ill repuke" where one might catch a "venerable disease." In the end, it turns out that the Commodore is Popeye's Pappy, who left home when he was two. As proof, they both pronounce alike, have bulging forearms, chew pipes, and are one-eyed; the Commodore also keeps an empty picture frame with "Me son" fondly scrawled inside. Fond memories follow the reunion: Pappy throwing the baby Popeye up in the air and not being there when he came down, presenting the child with an electric eel. Popeye has always detested spinach but at the climactic underwater fight, Bluto forces Popeye to eat it and the water starts bubbling because of the energy released and Bluto is bested.

Strange though it seems, there are aspects to the film that are reminiscent of *McCabe and Mrs Miller*, especially the picture of a small town in which might is right and the cowardly town folk who will not stand up to be counted in the time of a crisis, a motif originally to be found in Fred Zinnemann's western *High Noon* (1952), and their incapacity to stand up to authority. Popeye being the Commodore's long-lost son when the Commodore has been the political power behind Sweethaven is also a satirical turn.[17] Nothing in the film is as sunny as it seems and, in that way may be compared to *Thieves Like Us*, in which the warmly comic appearance hides violence underneath. One could say that the satire in the film comes out of making the world of the comic strip a metaphor for America and not only because of its anthem. Virtually every virtue that the people or space *do not possess* correspond to those held up publicly.

The Arthouse films

As already indicated, a significant part of Altman's satirical vision is manifested in generic subversions, and the logic is that genres preserve historical memory with the public by constructing myths around the historical moment. Subverting the genre is tantamount to interrogating historical memory and even *Popeye* does this. The character of Popeye is a memory of seafaring America represented by classics like Herman Melville's *Moby Dick* (1851) and Jack London's *The Sea Wolf* (1904). Walter Benjamin[18] finds in the figure of the trading seaman the embodiment of transmittable wisdom, acquired in traveling to distant lands and Popeye is nothing if not "wise." The cartoon series is about Popeye the returned sailor, but his world is shown only in terms of a few characters not of his sterling nature, like Wimpy and Bluto. His unschooled wisdom is ingenuous and this is given emphasis by his eccentric use of the English language. Altman creates a whole self-contained world drawn from this small level of detail and allegorizes America through it, the simpleton-hero's journey through the complications of the milieu, at the same time being ambivalent about the hero's relationship to power; the villain Bluto who is ultimately defeated is only Pappy's agent in Sweethaven.

But Altman also made films that are not from genres but realistic exercises depicting aspects of American society and I will examine three of them in this section: *Nashville* (1975) a film about 24 characters in Nashville, Tennessee, in a few days leading up to a fundraising event for an outsider running for President, *Three Women* (1977) a quirky film set in a Californian desert town, and *Shortcuts* (1993) which takes some of Raymond Carver's short stories and weaves them into a single tale. The latter two films are not based on satirical material but are rich in irony and, since they function best as character studies in set milieus, I will confine my attention to that aspect of the two films. *The Player* (1992), a satire on Hollywood, simply pokes amiable fun at its standards and notion of cinematic achievement but without much of a bite.

Nashville (1975)

This film deals with the country music scene in Nashville and is about various singers, their fans, and people with singing ambitions coming together in Nashville. The linking event is a fundraising event for presidential candidate Hal Phillip Walker of the Replacement Party, with an agenda that includes removing lawyers from Congress and replacing them with professionals from various vocations. We never see Walker but his vehicles are about relaying his words. The actors are mostly those who have been in his earlier films and many are also singers who wrote their own songs. The best way to describe the film is to begin with some of its *dramatis personae*:

HAVEN HAMILTON (HENRY GIBSON):	A country singer with political ambitions, manipulative and basically propagating right-wing values around the family and nation. His son Bud is a graduate of Harvard Law School who cannot sing but manages his father's business. His companion is Lady Pearl (Barbara Baxley), a great fan of the Kennedys, who manages a club and is usually inebriated.
DEL REESE (NED BEATTY):	Haven Hamilton's lawyer and also an organizer for Walker's campaign.
JOHN TRIPLETTE (MICHAEL MURPHY):	A smooth-talking consultant from California for Walker's campaign.
LINNEA REESE (LILY TOMLIN):	A gospel singer and Del Reese's wife, with two deaf children, whom she adores.
BARBARA JEAN (RONEE BLAKELY):	An emotionally fragile country singer, who is a star in Nashville but is on the verge of a collapse. She is managed by her calculating husband Barnett (Allen Garfield).
CONNIE WHITE (KAREN BLACK):	Barbara Jean's hated rival. The two never appear on stage together.
TOM FRANK (KEITH CARRADINE):	Member of the folk-rock trio Bill, Mary, and Tom and an inveterate womanizer. Among his conquests are Mary, who is Bill's wife, and Linnea Reese, whom he seduces.
OPAL (GERALDINE CHAPLIN):	A BBC reporter intruding everywhere and practicing highly opinionated pieces in private, in highly colored prose. Sleeps with Tom Frank and admits to a gathering that includes Mary—another of Tom Frank's conquests—that she "knows him in the Biblical sense."
MR GREEN (KEENAN WYNN):	His wife Martha is ill in hospital, on the same floor on which Barbara Jean

	is staying after her collapse. He runs a boarding house where another character Kenny Frasier stays. His niece LA Joan (Shelley Duvall) is visiting to see her aunt but is too busy trying to meet celebrities, and perhaps have sex with them.
PFC. GLENN KELLY (SCOTT GLEN):	A Vietnam War vet and a fan of Barbara Jean. His mother saved Barbara Jean from being burned to death and he spends his time attending the singer's scheduled events or in the hospital where she is interred.
WINIFRED (BARBARA HARRIS):	An aspiring singer who is running away from her husband; she hardly ever gets a chance to sing except at the climax.
TOMMY BROWN (TIMOTHY BROWN):	An African American folk singer who is interviewed by Opal.
SUELEEN GAY (GWEN WELLES):	A waitress at the airport lunch counter with great singing ambitions as a country singer but no talent at all.
WADE COOLEY (ROBERT DOQUI):	A cook at the airport lunch counter and friend of Sueleen who tries (unsuccessfully) to let her know that she has no talent. He publicly calls Tommy Brown the whitest nigger in America.
ELIOTT GOULD AND JULIE CHRISTIE:	The two stars (from *The Long Goodbye* and *McCabe and Mrs Miller*) appear as themselves passing through Nashville.

There are key moments in the film that deserve comment and three songs that border on self-parody are by Haven Hamilton. The first by Haven Hamilton celebrates the bicentennial ("We must be doing something right to last two hundred years").[19] Another song from Haven is purportedly from a man addressing his lover, asking her to unpack her bags.[20] There are three reasons why he cannot leave his wife, viz. his three children who he names and describes. Another great musical moment is Tom Frank singing a love song "I'm Easy," at least three women in the audience believing that the song is addressed personally to her. Barbara Jean rambling about her childhood when she has to sing to a restless audience and being hastily escorted away is another. There is a generally liberal atmosphere prevailing and Tom Frank asks Pfc. Glen Kelly if he "killed anyone this week." Opal of BBC tells the same Pfc. Glenn Kelly that she can tell by his face that he has been in Vietnam. Sueleen Gay's effort at holding an audience marks another high point when she sings songs composed by her but ends up coerced into a striptease. Wade Cooley, when he hears about it, tells Sueleen, "That's awful girl; they'll kill you in this town," and follows it up with advice to give up singing, but it has no effect. At the climax of the film Mr Green's mysterious lodger Kenny Frasier shoots at Barbara Jean and Haven Hamilton and the former is badly wounded. At this moment, Winifred shows presence of mind and steps in to sing and *Nashville* ends on this note.

Nashville, because of its scale and its incorporating so many subplots dealing with relationships about Americans of various types and also providing an outside view of them, has been seen as a eulogy of "national character," with an overarching commentary provided by the upstart candidate Hal Phillip Walker and his relayed talks on what is wrong with American politics:

> Altman sees the American as one whose spiritual home is not Washington, D. C., nor New York, and certainly not Hollywood, but the "heartland." The city of Nashville is an appropriate metaphor for that more than geographical region, for it is a sizeable urban center such as most Americans live in, but it perceives itself as still being in touch with the rural world outside its environs. Spiritual resident of Nashville, the American remains something of an innocent, identifying, not with the anti-heroes of existentialism (and of Robert Altman), nor with movie stars (like Julie Christie and Elliot Gould, who make cameo appearances as themselves in *Nashville*). Rather, the American's most revered "gods" and "heroes" are the Country-and-Western singers, with their down- home accents and their Cadillacs [...] Walker's voice, with its non-Eastern but otherwise unregionalized accent, proclaims a vaguely populist message, self-consciously naive but grounded in native American common sense.[21]

But from whatever has been said about the film, its satirical strain should be self-evident and some of it is unforgiving. As an instance of its irony is the dramatic commentary of the BBC reporter Opal at odds with what each situation merits. Altman has generally paid great attention to sketching out character and it is partly through his characterizations that he subverts generic myths. *Nashville* is all about characters and interactions, commenting on a variety of social matters, which should be evident from the description of the people in the film and the situations provided above. But a factor not noted hitherto is that few people undergo moral transformations in his films; no one learns lessons despite their errors being costly. They negotiate their ways out of situations or succumb to them, but remaining what they are, just as Popeye is "all that he is." If *McCabe and Mrs Miller* is a love story, the protagonists still find separate destinies, Mrs Miller in an opium haze while McCabe dies unsung a short distance away. In *Thieves Like Us*, Mattie, who is trusted by Bowie, sends him to his death when his ways endanger her family and business.

This sense of an open ending in Altman's films is not pessimism as much as the absence of a resolution. A resolution is the cessation of conflict and, in genre films, which have much to do with national history; the resolution is a reconciliation of the protagonist to the system, an optimistic or a pessimistic one. Polanski's *Chinatown* (1974) deals with the triumph of evil but that is a realization that comes to the protagonist, who cannot be the same again. It is the recognition of the small man's helplessness against economic power but is still reconciliation. Given the ubiquity of the reconciliation, what are we to make of Altman's "open endings" that negate it? My proposition is that the absence of a "reconciliation" in film after film points to Altman as an outsider, critiquing the milieu but neither hoping for change nor reconciled to it. In *M*A*S*H*, the protagonists are not transformed but are only moved out of the war on transfer. That is not a position anti-war films take but Altman takes the viewpoint of someone not hoping to change something.[22] The characters not coming together in *Nashville* despite the nation being the central issue but marching toward multiple destinies[23] is indicative of his being removed from it.

Three Women (1977)

The film, a departure for Altman in that it is more character driven than social study, was apparently based on a dream[24] he had had. The film is set in a small desert town in California; Pinky Rose (Sissy Spacek) from Texas comes to work in a health spa for elderly people and put in charge of an older girl Millie Lammoreaux (Shelley Duvall). Millie is a friendly sort always amiable through chattering, although few people respond to it and her remarks are ignored. Pinky is girlish and awkward and looks up to Millie as someone with great appeal. Since Millie's roommate has just left, she moves in with her into the Purple Sage Apartments, where again her neighbors ignore her, although she is forever trying to be amiable. The only other attraction in the town is where Millie hangs out, a tavern and shooting range named "Dodge City" owned by Edgar Hart and his pregnant wife Willie, who also owns Purple Sage. Edgar dresses like a cowboy, complete with a pistol, and is a former stunt double for a TV star. Willie is silent through most of the film and spends her time painting bizarre figures on the walls and the bottoms of swimming pools depicting beastlike humans with prominent breasts or genitals. Not many people come to Dodge City and those who come are mainly off-duty policemen. Edgar is the only one who responds to Millie but he is obviously no good.

Altman's use of his milieus has always been brilliantly evocative and *Three Women* stands out even among his films. Virtually everything is counterfeit around here, not least of all the cheer that Millie is radiating. Millie is always telling stories about her dates but we never see anyone showing interest—"I've stood him up several times but he keeps coming back for more," as she tells someone about someone else. A telling feature of the place is perhaps that the only men available to her, or her former roommate Diedre, are off-duty policemen.[25] Edgar is usually playing the fool pulling out his nickel-plated pistol and challenging the girls and Dodge City itself is fake, gallows complete with noose constructed outside and a rubber rattlesnake under a thermocol rock. At the Purple Sage Apartments, Millie is cheerfully promising to join the neighbors for a swim party or an eat-out but she is never actually invited. We often hear rude audible responses to her sunny greetings but she seems oblivious to it.

The high point of Millie's life is Diedre wanting to drop in with three "cute" men (also policemen) for a party and the two girls getting ready buying a whole lot of groceries to cook up a fine dinner, only for her guests to make excuses at the last minute, even preferring an outing in Dodge City. This makes Millie mad and she blames Pinky. Their apartment has been thoughtfully equipped with an extra roll-down bed in the living room if one of the girls finds a date, and Millie returns late that night with the only available man—a drunken Edgar. Before shutting herself in their room with him, Millie makes it clear to Pinkie that she will not have her ruining her fun. Pinkie is distraught and jumps into the pool from the first floor, seriously hurt but rescued by the neighbors and Willie, who sees Edgar coming out of the apartment with Millie.

This first part of the film gives us a brilliantly ironic sense of the milieu rarely equaled in American cinema but the second part is difficult to interpret—in the normal sense of what the audience can make of it. A primary irony lies in its characters having self-images at odds with the way we are made to see them. In the other films I have dealt with, Altman has never tried to get into people's heads and social behavior has been all that he is interested in but here he tries to probe psychology in the second half through Pinky's dream and engineering a crisis of identity between the girls. This distance from psychology is also true of the other filmmakers I have dealt with; confining their attention to the surfaces of things represented by the milieu or behavior makes their approach suited to satire. One cannot imagine Fox or Ali from Fassbinder's films having dreams; if Jack Torrance in Kubrick's *The Shining* has a nightmare (not pictured), it is as a portent of what is to come and not indicative of his psychology. It is the same with Millie and Pinkie, who are satirized as social beings and not as people with interiors.

The first part of *Three Women* is a sharply etched picture of tawdry postmodern America.[26] Millie has constructed her life according to the models set by lifestyle magazines and the elaborate preparation of the two girls for the failed party with Diedre and the two policemen is its high point.[27] Through the dream sequence and the subsequent ambiguities, Altman is negating his approach thus far and we are left perplexed without feeling obliged to interpret the film, although that is not to deny that psychoanalysts may still make something of the second part of *Three Women*.[28] There are indications that Altman was inspired by Ingmar Bergman's *Persona* (1966), but Bergman makes different kinds of films that are preoccupied with inner states. His films have also never given us such sharply etched pictures of the milieu as Altman does here.

Short Cuts (1993)

This film is based on nine short stories and a poem by Raymond Carver, a "minimalist" writer who mainly explored marriage and family issues in his stories. His stories are full of irony, which places his perspective close to Altman's although his stories are pictures of isolation while Altman's characters are gregarious. Carver situated his stories in the Pacific Northwest but Altman sets *Short Cuts* in Los Angeles. To give an idea of the kind of stories Carver wrote, "They Are Not Your Husband" is about a jobless alcoholic whose wife is a waitress in a café and he visits her when he is off-duty. One day, he sees a group of men in the café taking a peek under his wife's dress when she is bending down and passing disparaging comments about her "big ass." He pretends to be unrelated to his wife but decides to put his wife on a diet at home. This is one of the stories Altman uses in *Short Cuts* along with others centered usually on family relationships. The film is like *Nashville* in bringing together a host of characters to comment upon the milieu but, where they were all connected to the Country Music scene in some way or another in *Nashville*, the characters here are connected only in that they live in Los Angeles. The film also scrutinizes their private lives which *Nashville* did not do.

The film is set at the time of a medfly infestation and begins with Dr Ralph Wyman (Matthew Modine) and his wife, Marian (Julianne Moore), meeting another couple, Stuart (Fred Ward) and Claire Kane (Anne Archer)—an out-of-work salesman and a party clown, respectively—at Zoe Trainer's (Lori Singer) cello concert and make a spontaneous Sunday barbecue dinner date. Marian's sister, Sherri (Madeleine Stowe), is married to philandering cop Gene (Tim Robbins), who invents unbelievable stories to hide his affair with Betty Weathers (Frances McDormand). Betty is in the process of divorcing a helicopter pilot, Stormy (Peter Gallagher). Waitress Doreen Piggot (Lily Tomlin) is married to an alcoholic limo driver, Earl (Tom Waits). The film is so cut up into short bits that it is impossible to tell the story as a continuous whole. Following are some of the episodes in the film that illustrates the material.

Jerry Kaiser (Chris Penn), whose wife, Lois (Jennifer Jason Leigh), works from home as a phone sex operator, tending to the children while she talks to strange men; she has to fake orgasms, describe parts of the human anatomy in the process of getting her clients excited. Jerry is accustomed to it but later in the film he begins to show symptoms of jealousy that his wife never talks to him like this.

Stuart Archer and his friends go on a fishing trip, which involves hiking quite a distance. They camp on the edge of a river but, before commencing to fish, discover the nude body of a girl (later found to have been raped and murdered) floating in the river. Since it is already late and getting back immediately would be tiresome, they fasten the corpse to a tree to prevent it from floating away and continue fishing for another day before reporting it to the police. Anne Archer, when she hears the story, is disturbed and somehow holds Stuart culpable. But the trout caught on the excursion is taken by Stuart to the barbecue dinner with Ralph and Marian Wyman.

There are many more such stories and the general tone is more cynical than we have seen in Altman's earlier films. To give an idea of the tone Altman maintains, Marian Wyman is dressing for the barbecue dinner but spills wine on the front of her skirt and she takes it off to reveal herself naked below the waist; Altman keeps this up for some moments exposing her more than she need have been. At first glance there is little reason for this exposure—the woman might have worn panties—and it looks like unnecessary ridicule, although perhaps not titillation as some critics have seen it. But it can be read as a deliberate de-eroticization of marriage, as also is the incident involving the professional phone sex engaged in by Lois Kaiser.

Coming to Altman's transformation of the source material, Carver never wrote a novel and his stories are about inarticulate people stuck in ruts, alcohol being a frequent reason. The manifestation of the breakdown is most often in marriages and Carver is perhaps the poet of the marital breakup. By putting the stories together, Altman makes the film "about America" and socio-political as Carver's stories are not, since the recurring theme in Carver is the individual wrestling with his or her demons. The use of Los Angeles as the location is also significant since the city, being home to Hollywood, has a special place in the mythologies around the nation:

> The second question that I raised concerning setting is whether Los Angeles, "untapped" or not, can ever be a neutral setting. As Michael Sorkin (echoing many others) put it, "L.A. is probably the most mediated town in America, nearly unviewable save through the fictive scrim of its mythologizers."[29]

By using LA, Altman is representing America through one of its most globally visible cities and when he wanted to "take the roof off people's lives," his choice of LA as the site where this should happen is significant. Despite Altman's claim that he is not dealing with "Hollywood or Beverley Hills," everything in the film "has Hollywood written all over it."[30] By imposing an acidic vision on a variety of relationships constructed around monogamous heterosexuality, Altman is making a commentary on America. Marital infidelity and failing marriages are key motifs in *Short Cuts* while Hollywood steadfastly propagates the view that marriages as sacrosanct and affirms their significance to Americans in genres ranging from romantic comedies to noir. Noir stories, often situated in Los Angeles, frequently punish infidelity as in Billy Wilder's *Double Indemnity* (1944). When *Short Cuts* inspired Paul Thomas Anderson's *Magnolia* (1999),[31] that film affirms marital sanctity through Linda Partridge's (Julianne Moore) relationship with her dying husband. There is hardly such an affirmation of marriage in *Short Cuts*, teeming as it is with infidelity and strained marriages.

Altman's career had many ups and downs and after the stirring success of *M*A*S*H*, he had critical successes but few commercial ones. If he produced his best work in the 1970s, his films of that period, while satirical, are still more indulgent toward society and people than *Short Cuts*, which came immediately after his comeback film *The Player* (1992), which Altman described as a "mild film that offended no one."[32] *Short Cuts* shows Altman as a more bitter film-maker than in the 1970s. As indicated earlier, Altman may have seen himself as an outsider in America and he made films in Britain and Europe; and *Short Cuts*, which did not do well in the United States, won him the Golden Lion at Venice. Stanley Kubrick, we recollect, also lived in Britain and satirized the United States in his last films. Perhaps it takes the attitude of an outsider to systematically satirize aspects of one's own society or regard with deep irony most happenings around oneself.

I initially split Altman's into two categories: generic subversions and arthouse films. But both have an aspect in common: both explore national culture as an ideological construct with the subversion of mythmaking as their primary objective. The genre subversions, as I tried to demonstrate, were reworking mythologies connected with each genre category. The arthouse films only seem different since *Nashville* is focused on the people involved in a form of national culture and commonly partaking of it, while *Three Women* is about people trying to live the kind of life publicized as glamorous "lifestyles" in media mythmaking. *Short Cuts* is essentially an ironic take on the American marriage, debunking the mythology around it. Altman's films are not always satire, but when they are, it is the mythological constructs around the American nation that are his primary targets and, in debunking them, he strikes very deeply.

NOTES

1. Patrick McMahon, "Robert Altman's Cinema of Subversion," *Film Matters*, 2.1 (Spring 2011), 21.
2. A useful point of reference would be Roland Barthes who notes that myth is a kind of language and a set of conventions by which the exigencies of a historical moment are given eternal justification. Roland Barthes, "Myth Today," in *Mythologies* (London: Paladin, 1973), p. 143.
3. "(Altman) survived in uneasy alliance with the studios, somehow managing to make his own films his way." James Monaco, *How to Read a Film: The World of Movies, Media, and Multimedia* (New York: Oxford University Press, 2000), p. 353.
4. "At that time critics like Pauline Kael had a lot of influence, film started opening up more as an art form, and this was then closed off in the 1980s. Now, of course, everything is driven not by who makes the film but who's in it, will it make $100 million, will it work for fourteen-year-olds? The films I made didn't cost that much, the competition wasn't so heavy and the audiences were there." David Thompson (ed.), *Altman on Altman* (London: Faber and Faber, 2006), p. 70.
5. Thompson, p. 51.
6. Thompson, p. 74.
7. Robert Warshow, "Movie Chronicle: The Westerner," in *Film Theory and Criticism: Introductory Readings*, ed. by Gerald Mast and Marshall Cohen, 3rd edn (New York: Oxford University Press, 1985), pp. 435–36.
8. It has been noted that endlessly repeated but never challenged, stock thematic motifs encapsulate bits of folk wisdom and this is particularly so when the films deal with the past as the western does. Paul Loukides, "Taking Inventory: Stock Elements of American Feature Films," in *Symbiosis: Popular Culture and Other fields*, ed. by Ray R. Browne and Marshal W. Fishwick (Bowling Green: Bowling Green State University Popular Press, 1988), p. 55.
9. John G Cawelti, "*Chinatown* and Generic Transformation in Recent American Films," in *Film Theory and Criticism: Introductory Readings*, ed. by Gerald Mast and Marshall Cohen, 3rd edn, p. 503.
10. Stanley Cavell, "The World Viewed," in *Film Theory and Criticism: Introductory Readings*, ed. by Leo Braudy and Marshall Cohen (New York: Oxford University Press, 1999), pp. 337–42.
11. "He gave actors individual microphones, had them deliver lines randomly and talk over each other; he also allowed them to wander around the set or location, while he followed them surreptitiously with those long lenses as a zoologist might follow animals roaming around a cage. Weaving all these layers together in the editing suite, the result was like being far away from events yet eavesdropping on them." Mark Cousins, *The Story of Film* (London: Pavilion Books, 2004), pp. 336–37.
12. "While the police use every imaginable deception and draw on great reservoirs of men and machinery, Bonnie and Clyde succeed through their innocent playfulness, cunning, and

the aid of the dis-possessed migrant farmers. Common people, no matter how ordinary, ignorant and impoverished, can always unite to outwit those in power. The message is as old as the printed word. Like Tom Jones and Huck Finn, Clyde Barrow is uncivilized, uncouth, and uneducated, but his cleverness makes him typical of the American rogue hero and his predecessors." The villain in the film is a representative of the law, a Texas ranger, who pursues them with personal animosity. FD Williams, "The Morality of Bonnie and Clyde," *Journal of Popular Culture*, 4.1 (Summer 1970), 299–307.

13. The tension between the individual exceeding his freedom and the restricting state owes to the special importance given to "individuality." "Individuality" is best approached through the notion of "liberty" and American cinema is informed by a notion of individuality owing to the intellectual tradition of John Locke, Adam Smith, and John Stuart Mill. This tradition defines "liberty" and "freedom" largely in negative terms. It proposes a legal sphere in which individuals remain unobstructed by the external authorities of church and state. It has looked to laissez-faire capitalism for its fundamental values, emphasizing the independence of the individual—and above all the "rational" individual—from political, social, and economic constraints; this school celebrates the free pursuit of private gain in the open marketplace of material and cultural goods. Isaiah Berlin, *Four Essays on Liberty* (Oxford: Oxford University Press, 1969), pp. 123–30.
14. Walter Kerr, *The Silent Clowns* (New York: Knopf, 1975), pp. 14–31.
15. Michael Barrier, *Hollywood Cartoons: American Animation in Its Golden Age* (New York: Oxford University Press, 1999), p. 3.
16. "In *A Clean Shaven Man*, released in February 1936, Popeye pulls a barber chair out of the floor and spins the chair's seat off its base, so that it whirls through the air and forces Bluto against a wall. Then the seat, with Bluto in it, bounces back onto the base of the chair, which Popeye is holding. There is no sense that this might somehow be physically possible; the seat simply floats. The slow and even Fleischer timing, sometimes so effective in earlier films at contributing to a dreamlike atmosphere, was much more problematic in cartoons organized around coherent narratives and dependent on audience interest in their protagonist." Michael Barrier, *Hollywood Cartoons: American Animation in Its Golden Age*, p. 186.
17. National heroes being related by blood to people with dubious connections in politics as well as the underworld is not uncommon. President Kennedy was a son of Joseph Kennedy who had ostensibly made part of his enormous fortune through bootlegging in the prohibition era. Phillip Whitehead, "The Bootleg Politician," *The Independent*, October 11, 1992. https://www.independent.co.uk/arts-entertainment/the-bootleg-politician-he-could-have-anything-he-wanted-except-the-thing-he-wanted-most-so-joe-1556722.html. Accessed July 3, 2020.
18. Walter Benjamin, "The Storyteller," in *Illuminations*, ed. by Hannah Arendt, trans. by Harry Zohn (London: Jonathan Cape, 1970), pp. 84–85.
19. My mother's people came by ship / And fought at Bunker Hill / My daddy lost a leg in France / I have his medal still / My brother served with Patton / I saw action in Algiers Oh we must be doin' somethin' right / To last 200 years.

20. Unpack your bags, / And try not to cry, I can't leave my wife, / There's three reasons why, There's Jimmy, and Cathy, and sweet Lorilei, For the sake of the children, / We must say goodbye.
21. Roger B Rollin, "Robert Altman's *Nashville*: Popular Film and the American Character," *South Atlantic Bulletin*, 42.4 (1977), 42–43.
22. This is not a position he could have taken with the Vietnam War, still in progress in 1970, and hence opts for Korea. He used the anti-military sentiment generated by a contemporary war to make an anti-war comedy about a past one, a war relegated to forgotten history. That war was so distant that Altman's film does not even name the enemy.
23. The characters come together nominally in the culminating concert but the common realization that diverse people attain in films about collective endeavor around the nation—as in a war film or William Wyler's *The Best Years of Our Lives* (1946)—is not in evidence. They are still speaking in different voices; even two friends like Sueleen and Wade are not with a single understanding, when she is only amused over his warning. LA Joan also never visits her dying aunt, although she is in Nashville only for that.
24. Mitchel Zuckoff, *Robert Altman: The Oral Biography* (New York: Vintage, 2010), p. 320.
25. One wonders why this detail provides such a telling aspect of the milieu. This is probably because policemen are "persons" only off-duty. The impersonality demanded of a policeman in his work implies that. On-duty, they are dealing with people as impersonal instruments of the state. When a girl dates a policeman, she, in effect, only dates half a person. Diedre or Millie's dates all being policemen say something about the social life in the town.
26. The nostalgia mode as represented by Dodge City is a characteristic of postmodernism. Frederic Jameson, "Postmodernism and Consumer Vulture," in *Postmodernism and Its discontents: Theories, Practices*, ed. by E Anne Kaplan (London: Verso, 1988), pp. 18–20.
27. Roland Barthes's examination of the cookery columns in the women's magazine *Elle* in his essay "Ornamental Cookery" could be an interesting juxtaposition here. Barthes notes that the kind of recipes published in a magazine meant predominantly for working-class women would be unaffordable to them. Roland Barthes (ed.), "Ornamental Cookery," in *Mythologies* (New York: The Noonday Press, 1972), pp. 78–80.
28. Glen O Gabbard and Krin Gabbard, *Psychiatry and the Cinema* (Washington, DC: American Psychiatric Association, 1999), p. 223.
29. Kasia Boddy, "Short Cuts and Long Shots: Raymond Carver's Stories and Robert Altman's Film," *Journal of American Studies*, 34.1 (2000), 13.
30. Michael Wood, "Why the Birthday Party Didn't Happen," *London Review of Books*, March 10, 1994, p. 19.
31. Patrick McMahon, "Robert Altman's Cinema of Subversion," *Film Matters*, 2.1 (Spring 2011), 24.
32. DVD commentary on *The Player.*

5

Paul Verhoeven: Satirical Impulse Goes Clandestine

Paul Verhoeven is different from the other filmmakers in having usually been unabashedly commercial and with a number of megahits to his credit. He has won best film awards several times both in his home country Holland and the United States and has also won the Golden Raspberry for both the worst picture and worst director (*Showgirls*, 1995). He began working in his home country but moved to Hollywood later to make films in a variety of genres, including science fiction (*Starship Troopers*, 1997) futuristic fiction (*RoboCop*, 1987), and neo-noir (*Basic Instinct*, 1992). His films are notorious for their graphic sex and violence but it is recognized that there are elements of satire in his films, often disguised as sensationalism. His American films have been written about but it is difficult to find academic enquiries into his Dutch films like *Turkish Delight* (1973), although they are hugely popular in Holland. I will divide this essay into two parts, beginning by looking only at *Turkish Delight* among his Dutch films and then at his key films in the United States, where I believe his Dutch background is relevant.

The Dutch period

Verhoeven's films are known for their explicit sex and violence and it would seem that he comes from a milieu with a degree of license in such matters, unimaginable in most other countries. Jan Wolkers's novel *Turks Fruit* (1969), on which the film is based, opens with the unnamed narrator masturbating over nude photos of his departed lover, Olga, and rereading her old letters: "While I'm writing this to you my cunt is making sucking motions, like a baby's mouth," reads one, and this novel was on the school syllabus, a musical even being made with it.[1]

Amsterdam has been known the world over for its hash bars or "coffee shops," where patrons can "legally" purchase marijuana and hash in small amounts; Quentin Tarantino's *Pulp Fiction* (1994), in fact, celebrates it. In the last 40 years, the city has become known for its tolerance of prostitution and the use of cannabis.

But this permissiveness is not a phenomenon of the twentieth century and the Netherlands has a history of hedonism among youth that goes back to its Golden Age in the late sixteenth and seventeenth centuries, when artists like Rembrandt and Vermeer were working.

In the first place, religion was less oppressive in the Netherlands. With the eclipse of the Roman Catholic Church in the Dutch Republic, there was no predominant religion to take its place. The Protestant Church tried to gain the former's position as the state church but failed. The religious landscape of the new country was fragmented and divided among different religions, which lived side by side and were obliged to tolerate each other.[2] Second, Amsterdam had become the richest city in the world. After the fall of Antwerp in 1585 to the Spanish forces in the Eighty Years' War, commercial trade in northern Europe was diverted overnight from Antwerp to Amsterdam. From one day to the next, Amsterdam became the new staple market of Europe. The sealing off of Antwerp's harbor created a large influx of immigrants, who fled the city and resettled in the north. They brought their knowledge, money, and trade contacts with them. The influx of immigrants from Southern Netherlands unleashed a wave of economic prosperity and a building boom and generated unprecedented employment.[3]

The Dutch sociologist Henk Becker, in his book *Generations and Their Chances* (1992), proposes that young people growing up during economic hardship and crisis, war, or times of disease are more likely to be pessimistic and have a negative view of the future. Young people who grow up during times of economic prosperity and peace, by contrast, have an optimistic outlook on life.[4] In the early seventeenth century, there was social turmoil in parts of present-day Germany, where a bloody conflict was fought between Roman Catholics and Protestants. During the Thirty Years' War (1618–48), entire towns and villages were destroyed. The countryside was looted, crops were destroyed, and livestock were indiscriminately slaughtered. If the inhabitants did not perish by the sword, they were vulnerable to famine and plague. There was little future for the young people in these regions. For them, moving to the flourishing cities of the Northern Netherlands was often the only option to build a new life.[5]

Given these conditions, many people had great amounts of leisure. Unlike in contemporary society, where free time is equated with wealth and consumption, in the Dutch Golden Age it meant that many people had nothing to do for several days. When young people were bored, they were more likely to commit acts of violence, obsess about their appearance, drink too much alcohol, fornicate, and take drugs. During every period of economic prosperity, these aspects of youth culture reappear, but they often assume a different shape or form. In the Dutch Golden Age, juvenile violence was rampant. Besides violence and hedonism, parents were concerned that their children were drinking too much alcohol.

Young people in the seventeenth century were notorious for their binge drinking. The young people of the Golden Age were the first generation in history to embrace smoking and turn it into a mainstream habit and national pastime.[6] The Netherlands did not generally suffer in history although Napoleon occupied it and made his brother king in 1806. In World War I, it was neutral while in World War II, Hitler overran it in four days. This is the milieu Verhoeven comes from and the explicit sex and violence in his films should be seen in this light, as the cultural product of a liberal society that had generally enjoyed prosperity.

Turkish Delight (1973)

Turkish Delight is not a film that can be readily recognized as "satire" since it is a love story and does not scrutinize Dutch society. But there are also covert ways in which a society can be satirized and Robert Altman did that by looking at marital relationships in *Short Cuts*; Kubrick ostensibly produced satire in *The Shining* in telling the story of a family isolated by physical circumstances. The genre of Menippean satire, as I noted in the introduction, attacks mental attitudes rather than specific individuals or entities and mental attitudes are the products of a structured society.

Turkish Delight begins with Eric Vonk (Rutger Hauer), a sculptor, asleep in bed in his dirty studio, his genitals exposed, dreaming of killing a man and a woman. He wakes up, masturbates before the nude picture of a woman, and then sets out looking for random women and having sex. Eric Vonk is a Bohemian and lives for each moment; so complete is his sense of existential freedom. But Eric is evidently in a state of distress and this is because of his estrangement from Olga (Monique van de Ven) whose nude statue brings back memories; the film then embarks on an elaborate flashback from two years before in Valkenburg.

Eric is still an apprentice sculptor and he creates a minor scandal by sculpting Lazarus covered with maggots and worms since he was dead four days before Jesus raised him. Later at a dinner with the municipal authorities, he makes a ruckus over the eye of a game animal he found in the meat, and the other young artists follow his cue. The authorities are evidently too ineffectual to contain Eric's mischief. After this event, Eric runs into Olga, when he thumbs a ride from her and gets her involved in an accident.

Verhoeven's method of filmmaking is not understated and subtle but just the opposite; Eric is perpetually flirting with death and the accident in which Olga is injured comes as no surprise. One could examine this aspect of Eric in terms of Eros and Thanatos or the life and death instincts[7] but Verhoeven does not try to get into Eric's mind. We do not speculate on why he is behaving. After the accident, which is serious, Eric senses a soulmate in Olga and tries to pursue her. Her parents own a large store and he tries visiting her there but is evicted, and he is evidently not wanted. But here again, we do not see their opposition to him as representing "bourgeois opposition" to true freedom as we might have in another context. Michel Poiccard's self-destructive Bohemianism in Jean-Luc Godard's *Breathless* (1960), for instance, could be regarded this way, as social rebellion, but Eric Vonk's conduct does not emerge that way. Her mother, for instance, has usually a half-smile on her face even as she is shooing him away, implying that he is not completely an undesirable. In any case, Eric runs into Olga at a fairground and they resume their relationship. When the two are caught together by her mother, it is revealed that Olga is a minor but he still proposes marriage and, with her father on his side, the parents come around to Eric and Olga duly becoming husband and wife.

Eric and Olga's romps continue after their marriage but there is death and decay lurking around even in their pranks. Eric, for instance, makes fun of Olga's mother when he puts an inflated balloon inside a bra of hers and the fact is that she has only one breast, having recovered from breast cancer and mastectomy. Another scene shows them visiting a surgeon friend Paul who is supervising a patient's brain scan. The film is still going through with the explicit sex but the undercurrent of death is unmistakable. There is, for instance, a sequence of Olga defecating and coming out frightened, seeing blood in her feces but Eric, after physically scrutinizing it, recognizes traces of last evening's dinner, which included beet. This association of eroticism and defecation[8] was made by satirist Jonathan Swift, although Verhoeven does not maintain a bitter Swiftian tone, but a more genial one through most of the film. But just when he is covering her in wildflowers ("Persephone," his next project),[9] he receives news that her father is seriously ill. When the flowers are pulled away, they leave behind worms and dead leaves. We are not told what is wrong with the old man but he cannot control his excretory functions. But even as her father is dying Eric and Olga have sex.

After her father's death, Eric is expected to look after the family store but the two return to Amsterdam. She works in a milk bottling plant and he manages to sell a picture of the two of them in the sex act, which angers her. A while later at a party with the family, he spies her kissing a stranger and responds by spewing vomit on Olga and her mother and following it up by slapping Olga. This is what led to the breakup and we are now back in the present with Eric, alone.

The last part of the film deals with their divorce, Olga marrying an American but coming away after an unsuccessful year at being wedded to someone else. Eric still yearns for her and when she is sleeping, has sex with her, which amounts to rape. But it is discovered that she has a tumor in her brain and an operation is unable to take it out fully. Her "cranium" is clean-shaven with a small square window on it and Eric gives her a red wig to wear. But she dies and the last image is of Eric disposing of the wig in a garbage dump, where it is swallowed up by an excavator.

If one were to search for a word to describe the film, the word would not be erotic or romantic as much as "nihilistic." *Turkish Delight* appeared three years after *Love Story* (1970), a genteel story of love ending in the terminal illness and death of one of the lovers. But where that film was about "spiritual love" which is eternal since the spirit does not perish, Verhoeven's is about the flesh. The "tragic" conclusion of *Turkish delight* is therefore savage in its bluntness. Satire thrives on the pretenses and blindness of social life; our claim upon immortality—through religion and art—may be our greatest blindness since the "void" is something we cannot countenance.

When we look back at Dutch society of the fifteenth to seventeenth century, we find artists like Hieronymus Bosch (1450–1516) and Pieter Brueghel (1525–69) living in a rich society or being personally prosperous but producing among the darkest of artistic works, full of visions of death and hell as in Bosch's *The Garden of the Earthly Delights* (1495–1505) or Brueghel's *The Triumph of Death* (1562) in which pleasure and horror are satirically juxtaposed. It is irony or satire because it is debunking earthly pleasures, arguably an off-shoot of the social circumstances marking the Dutch society of its times. Paul Verhoeven does not belong to a religious era and hardly invokes hell but *Turkish Delight* continually deals with death. It allows freedom of purpose, but with the rider, death waits in every direction.

The American films

Where Verhoeven's Dutch films were, by and large, arthouse exercises, his American films are fantasies and sci-fi mainly and unabashedly commercial ventures. But even as they use sex and violence to sell themselves, they also have a strong satirical element, or, to phrase that more accurately, can be read more easily as satire and irony. Looking at this differently, one could say that in his Dutch films, he is dealing directly with aspects of the milieu including its past but he evidently cannot do that "authentically" in the United States since the milieu is not his own. European filmmakers have dealt with the United States, trying to give a different picture—often a more satirical one—than the one we get from Hollywood and a director would be Werner Herzog and I would cite his *Stroszek* (1977).

Verhoeven has a natural propensity for irony but, rather than give an outsider's picture of America, he gives us satirical pictures of the milieu but as refracted through the media—advertisements, TV shows, including news, and formulae from existing genres. Usually, this means over-the-top action and sleazy sex, which can be read either straight as exploitation or as a parody of exploitation films.

Verhoeven satirizes America largely by parodying media representations of it and his first Hollywood film *RoboCop* (1987) demonstrates this very well.

RoboCop (1987)

The film is set in Detroit in the near future where street violence is rampant, policemen being killed, and if a model is to be identified it would be Michael Winner's *Death Wish* (1974) with Charles Bronson in the lead role of a vigilante. The city in *RoboCop* has just signed a deal with the megacorporation Omni Consumer Products (OCP) to run Detroit's police department in exchange for letting OCP rebuild the run-down sections of the city into a high-end utopia called Delta City. Senior Vice President Dick Jones (Ronny Cox) demonstrates a new law enforcement robot, ED-209, which unfortunately malfunctions at a critical moment. An executive is asked to volunteer, point a gun at the robot, which promptly asks him to throw down the weapon in twenty seconds. He does so at Dick Jones's behest but ED-209 is not placated and starts firing its advanced weaponry, driving the OCP executives helter-skelter and killing the volunteer. ED-209 evidently needs to be reworked to make it effective. At this moment, Bob Morton, an ambitious executive, introduces his own experimental cyborg design, RoboCop. The chairman is dismayed at ED-209's failure and approves Morton's plan, much to Jones's ire.

At around the same time, Alex Murphy (Peter Weller) joins the Metro West precinct and becomes partner to Anne Lewis (Nancy Allen) and the two pursue a gang led by Clarence Boddicker (Kurtwood Smith) after an armed robbery. Investigating their hideout, an abandoned steel mill, Murphy kills one of the gang members but is ambushed by the leader, Clarence Boddicker. Boddicker and his accomplices shoot Murphy repeatedly with shotguns until Boddicker shoots him through the head. Anne arrives too late to save him. To catch Verhoeven's parodic tone, it is necessary to relate a happening pertaining to why Anne Lewis is delayed. She catches an African American member of the gang urinating against a wall and holds him up, making him turn around and his penis is apparently exposed. Lewis is intent on arresting him but cannot resist a downward glance, and the gang member pounces on her and disarms her. This is evidently not something that might happen in a standard Hollywood action film, which would take the episode very seriously, especially since a duty-conscious policeman has been killed.

OCP claims Murphy's body and turns it into RoboCop. The cyborg is programmed with three main directives: serve the public trust, protect the innocent, and uphold the law. RoboCop is assigned to Metro West, where he begins a brutally efficient campaign against crime. But although Murphy's memory was wiped, RoboCop begins to remember scenes from Murphy's life, including his death. His movements also mimic some of Murphy's mannerisms. Lewis also confirms to RoboCop that he is Murphy, much to his shock. On patrol, RoboCop foils an armed robbery perpetrated by Emil Antonowsky, one of the gang members who participated in Murphy's killing. RoboCop uses the police database to identify the gang members and also locate his home, now abandoned. Meanwhile, at the behest of Jones, Boddicker murders Morton. This, too, is over-the-top: Morton with a couple of overdressed, slatternly women declaring that he "loves to be with intelligent women," and also that "a mind is a terrible thing to waste."

Verhoeven crowds the film with details of the milieu pertinent to the Reagan era: like a Lee Iacoca Elementary School. Iacoca was an American automobile executive best known for reviving the Chrysler Corporation as its CEO in the 1980s, and Detroit is the home of the American automobile. The Reagan era was when the Star Wars program and the Neutron Bomb (which kills people but spares property) was news and a channel announces that the South African government is planning to end riots using a French-made Neutron Bomb. The channel also announces the President's first press conference from the "Star Wars Orbiting Peace Platform," where there is a power glitch, and the dignitaries float around weightlessly. A satellite mistakenly sends a laser beam to Santa Barbara, killing 113 people including two ex-presidents. Children play a table game called "Nuk-em" which concludes with a table-sized mushroom cloud. Delta City advertises itself through the slogan "The future has a silver lining" implying that the future is actually clouded over, except for the space of the elite. Japan was experiencing a boom period then and a surgeon endorsed a Yamaha mechanical heart.

RoboCop locates Boddicker at a cocaine factory, where he kills the workers present and captures Boddicker. But before RoboCop can kill him, Boddicker reveals that he is in Jones's employ. However, when RoboCop attempts to arrest Jones, RoboCop unwittingly activates a previously unseen fourth directive, implanted by Jones himself: "Any attempt to arrest a senior officer of OCP results in shutdown." RoboCop is a "product" and product-violation cannot be allowed. Jones admits his culpability in Morton's death and tries to kill RoboCop with an ED-209 unit but ED-209's three-toed feet will not allow it to climb down a staircase. The sight of the fearsome mechanical monster trying hesitantly to align its toes to the steps is hugely incongruous. At the climax of the film, RoboCop has Jones in his sights but Jones takes the chairman hostage and puts a gun to his head. RoboCop cannot act since Jones is a senior officer of OCP. This dilemma is however solved when the chairman, with a gun to his head, fires Jones, which means that he loses his "senior officer" status and RoboCop shoots him.

There are two aspects of *RoboCop* that deserve special comment here and one is that while Jones is the villain, the others are also lampooned, as for instance Morton. But the final resolution is achieved by Jones's elimination alone. What this suggests is that corporate is there to stay, as is its privatization of the police force, and the victory over Jones and his machinations does not amount to an actual victory. The unassailability of the corporate evidently echoes Altman's treatment of the mining company in *McCabe and Mrs Miller*. Even in *Popeye* it is only Bluto and not the Commodore ("Pappy") who is defeated and made to flee. *RoboCop* began with OCP designated the villain but ended with culpability transferred to one of its employees, a commonplace happening in the capitalist system where the corporate itself often remains immune to action. The film emerges as a sharp critique of the neoliberal resurgence in law and politics at the end of the twentieth century. In particular, it critiques the privatization of police services, and the expansion of private property claims to cover a cyborg policeman, regardless of its human components.[10]

The second aspect is the film's use of television as a ubiquitous presence. Its ironic use lifts us from the action narrative into a commentary on corporate control over society and how the electronic media manages public opinion. "You give us two minutes: we'll give you the world" intones the news show. An event covered is a murder-kidnapping by a former city councilman, holding the mayor and other hostages in city hall under siege. Before the cyborg solves the problem, we hear the paranoid man's demands, including a new car, a better one than the city gave him before. The man shouts out his idea of a luxury vehicle: "I want a car that gets really shitty gas mileage!" We have been seeing television messaging and this could be its effect on a gullible public.[11]

Total Recall (1990)

RoboCop worked perfectly as satire but Verhoeven is essentially producing mass entertainment in America and it may not be easy to fully reconcile the satirical impulse with thrilling climaxes and satisfying resolutions that are essential—given the large budgets he is working with and the commercial demands that his films need to meet. In *Total Recall*, for instance, there are too many twists for us to explain it entirely as satire. As already indicated, Verhoeven uses "excess" as an authorial strategy, which means one cannot look to his films for subtleties; the satirical effects are not in quiet inserts but in the broadest or loudest segments.

In 2084, construction worker Douglas Quaid (Arnold Schwarzenegger) is having troubling dreams about Mars and a mysterious woman there. His wife Lori (Sharon Stone) dismisses the dreams and discourages him from thinking about Mars where the governor, Vilos Cohaagen (Ronny Cox), is fighting a rebellion; the mineral Terbinium is mentioned. On the subway, he sees an advertisement from a company called Rekall, offering vacations. Or, to be more precise, it is offering memories of vacations—as implants. Quaid discusses this with a colleague Harry who warns him against it since "they mess with your head" but he is still tempted and visits their office. He wants a trip to Mars and they offer another addition, either to go in his own identity or another and he chooses to go as a secret agent and meet the brunette from his dreams. Unfortunately, when they are putting in the implant, something goes wrong and it turns out that Quaid already has a memory of Mars because he has been there as an actual secret agent, although most of it is clouded over. Rekall therefore erases the memory and sends him home. But on the way, he is attacked by people including his friend Harry and he kills them. At home, he is ambushed by his wife Lori who reveals that she is not his wife at all and that their marriage is a memory implant. Lori is actually the wife of Richter (Michael Ironside), Cohaagen's right-hand man. After evading his attackers, Quaid is left a suitcase containing money, gadgets, fake IDs, and a video recording. The video is of Quaid himself, who identifies himself as Hauser and explains that he used to work for Cohaagen, but switched sides after learning about an alien artifact on Mars and underwent the memory wipe to protect himself.

Sci-fi films are normally solemn affairs with messages and *Avatar* (which came later) bears a resemblance to *Total Recall* in being about mining a rare mineral on another planet, and the protagonist having to fight a human villain unsympathetic to the local cause. The local cause in *Avatar* was the environment the natives were fighting to save, while in *Total Recall*, the rebels are fighting for air. *Total Recall* has two different parts and while the first deals with the issue of the implanted memory, the second has to do with the struggle on Mars with Quaid/Hauser joining the rebels. The second has only a few satirical or parodic bits when the rebels are shown to be mutants created when the dome above their area was faulty and there is a parody of films like *Star Wars* in which no two creatures are alike. In "Venusville," where Quaid looks for the brunette of his dreams—named Melina—he is also introduced to a stripper with three breasts.

It is the implanted memory aspect of *Total Recall* that really functions well as satire. When the film begins and Quaid is trying to persuade Lori that they should move to Mars, he says he feels he was meant for "something more" than the life he is leading and it turns out that he is actually a secret agent with his memory erased. On the metro, he sees the ad from Rekall, which goes "For the memory of a lifetime: Rekall, Rekall, Rekall," in an ethereal song the way advertisements are done today. At Rekall, the receptionist is busy with her fingernails; they are blue and she experimenting electronically altering it to red and black to blue; she behaves flirtatiously as one might expect with a travel agent's office where they might sell you a vacation. The executive gives him all the sales talk such an executive might impart but offering "first-class memories" instead of a "first-class holiday." All the sights of Mars are mentioned including "Venusville" in a mischievous tone since Venusville is the sleazy quarter. He draws his attention to a real holiday being a "pain in the butt": lost luggage, lousy weather, crooked taxi drivers. A memory could be tailored to be perfect, no unpleasant experiences. The high point of the presentation is this question: "What is exactly the same in every vacation you have ever taken?" The answer is that "you" are the same and this can be remedied if you travel with another identity: a playboy, a billionaire, or a secret agent are those on offer. Taking a holiday from yourself is the latest thing in travel and it is called the "Ego Trip." Going as a secret agent, the promise is, "You will win the girl, kill the bad guys and save the planet." If one were to ask what makes this segment profoundly satirical, it is the sense that regardless of technological advances, business strategies remain the same, especially its methods of consumer persuasion. Where visionaries like HG Wells imagined humankind as transformed by future technological advances,[12] Verhoeven sees it as unchanging since it is created and maintained by the market, which mediates between humankind and technology; regardless of the advances, it suits the market to keep human nature predictably the same.

Basic Instinct (1992)

Verhoeven working in America is evidently beset by contrary impulses, the first to make films that will win him audiences and the second to follow his natural inclinations toward satire and, if the second one gets out of hand, he risks losing his audiences. By the time he made *Basic Instinct*, he perhaps came to the recognition that Americans—as also much of the West—were more squeamish about sex than the Dutch and he could draw audiences by playing up the sex while appearing to be dead earnest in his storylines. No film demonstrates this as clearly as *Basic Instinct* in which the satirical humor of his earlier American films is visibly absent. The film is generally described as a "psychological neo-noir thriller" and this is strange, considering that Verhoeven had not shown an interest in psychology in the past.

The film begins with a steamy sex scene involving a man, his hands tied to a bed, and a blond woman whose face we do not see. At the climactic moment, the woman grabs an ice pick and stabs the man repeatedly to death. It is San Francisco and homicide detective Nick Curran (Michael Douglas) investigates the murder of retired rock-and-roll star Johnny Boz, who has been brutally killed during sex with a mysterious blond woman. Nick's only suspect is Boz's bisexual girlfriend, crime novelist Catherine Tramell (Sharon Stone), who has previously authored a novel that anticipates the crime as it took place. Curran has anger management problems and was once involved in the accidental shooting of two tourists when high on cocaine and he is now required to report for counselling sessions to police psychologist Dr Beth Garner (Jeanne Tripplehorn), with whom he has been having a sexual relationship off and on.

A key motif in the film is "psychology" and both Catherine Tramell and Beth Garner studied it at Berkeley. Nick, who is now involved with Catherine Tramell sexually as well, discovers that Catherine is basing the protagonist of her latest book on him, wherein the character is murdered after falling for the wrong woman. Tramell has had access to Nick's psychiatric files through Lt. Marty Nielsen of Internal Affairs. There are other clues and it is learned that Catherine Tramell's (and Beth Garner's) professor at Berkeley was similarly done to death with an ice pick. Much is made of psychological profiling as providing the solutions but none of it gets very far. Sharon Stone, in fact, plays Catherine Tramell as impenetrable. Both she and Nick (after his "accidental" killing of two tourists) get through lie-detector tests with ease. She herself seems so opaque as to be outside the scope of psychological investigation. Eventually, it turns out that Beth Garner and Catherine Tramell had a lesbian relationship in college but both claim that the other had an obsession with her and there is no knowing which one is telling the truth. There are other murders and the crimes are eventually pinned to Elizabeth Garner. But at the closure, when Nick and Catherine are having sex, there is a hint that Catherine is intending to kill him with an ice pick, although that does not materialize. This has had critics expressing dissatisfaction that having two solutions to a murder mystery is ludicrous.[13]

The film has tried to elicit psychological interpretation and here is a woman psychologist's oppositional reading of *Basic Instinct*:

> *Basic Instinct* is a movie that promulgates a hateful message with such artful means that it provoked me to hate myself for loving it. It reminds me of seeing Triumph of the Will or Intolerance or Day of the Locust in that regard. It is clearly a part of the battle against women that Susan Faludi describes in *Backlash*. The story of the movie is a relentless attack on women. Educated women, this movie tells us, are dangerous, bloodthirsty lesbians. Even the ones who seem to love men are at least unethical and possibly murderous. The most attractive and openly sexual ones like Catherine, are the most alluring and the most deadly. Powerful women and women who achieve things, like writing books or getting a degree in psychology, are the worst.[14]

The film has also been studied from its syntagmatic structure and the following are the alternatives proposed as solutions[15]:

(a) The "writer" (Catherine) is a devious diabolical mind who wrote her novel as an alibi years before committing the murder.
(b) Someone (Beth) is so obsessed with the writer that she is willing to kill innocents in order to blame the writer. In a trick ending, the filmic apparatus undoes at the end the very solution it had proposed and maintains the ambiguity of its "whodunit."
(c) Instability of all three main protagonists Catherine, Beth, and Nick influencing the relationships as well as the solution.

My own reading of the film is contrary to all the above and I see the "ambiguities" not as pointing to psychological complexities but to a debunking of the psychological resolutions of the American crime fiction film. American films (*The Silence of the Lambs*, 1991) take the psychoanalytical method in crime solving more solemnly than those from, say, Europe, and Verhoeven is parodying the tendency. Why else would he show both Beth and Catherine naked except to help us gauge from their bodies as to which woman is the actual killer, given that he has shown us the actual murder while keeping the killer's face covered?

The "puzzle" in the film, in fact, has no solution but only pretends to have one. In the normal whodunit, we are presented with clues that we are required to follow to solve the puzzle but, here, Verhoeven is not leading us to a solution through clues; he is, instead, showing us the actual killing and asking us to identify the killer by looking at her body, also naked, and shown separately to facilitate it. He is riding on the moral squeamishness of audiences—that might hesitate to do this. It would, I propose, be seen as "honourable" by most spectators to identify the killer only by her face—even when she is naked and only her face is hidden! But the sequences involving the naked women are so glamorized that one overlooks the physical "details" and that is what Verhoeven is teasing us with, not intellectual solutions. The film is satire or parody that does not present itself as satire or parody—by being so straight-faced in every motif. I would take the utterance of a policeman—when Beth's hand in the killings is determined—that "You can't tell about people," as a hint that the psychological solution is even laughable.

Showgirls (1995)

With the phenomenal success of *Basic Instinct*, Verhoeven got a big budget to make *Showgirls*, also full of nudity and sex. But there is a difference in Verhoeven's handling of sex in the two films. *Basic Instinct* was deeply voyeuristic, although that was justified by the theme basically being an examination of the private doings of people and the secrets they hide. *Showgirls* is about a young woman dancer making a career in Las Vegas. Being about life played out under the arc lights of show business, it scarcely contains a private moment, which virtually rules out eroticism and the voyeuristic gaze. The absence of the voyeuristic gaze perhaps meant that despite its sensationalist theme focusing on a stripper and the enormous amount of nudity on offer, it did not do well either critically[16] or commercially, but was a smash hit in the DVD circuit suggesting that the people who watched it did not want to be caught at it!

In the film, Nomi Malone (Elizabeth Berkley) is a young hitchhiker who goes to Las Vegas hoping to become a dancer. After being robbed by her driver, Nomi meets Molly Abrams (Gina Ravera), a costume designer who becomes her roommate. Molly invites Nomi backstage at *Goddess*, the Stardust Casino show where she works, to meet Cristal Connors (Gina Gershon), the diva star of the casino's topless dance revue. Nomi telling Cristal that she dances at Cheetah's Topless Club has the latter deriding what she does: that it is not "dancing" and the insinuation is that it is prostitution. When Nomi is too upset to work that night, Molly takes her dancing at The Crave Club. Nomi is arrested after causing a fight involving James (Glenn Plummer), a bouncer. James bails Nomi out, but she then ignores him. James is African American like Molly, dances beautifully and thinks Nomi has talent as well. Cristal and Zack Carey (Kyle MacLachlan), the entertainment director at the Stardust, visit Cheetah's and request a lap dance from Nomi. Although the bisexual Cristal is attracted to Nomi, there is also an effort on her part to put Nomi in her place. Nomi reluctantly performs the lap dance after Cristal offers her $500, Cristal watching. James is at the strip club as well and witnesses the lap dance. He visits Nomi's trailer the next morning and chides her for what she is doing.

What *Showgirls* does *not* offer is a candid, behind-the-scenes look at the Las Vegas showbiz industry. As already indicated, Verhoeven's American films look at aspects of America but only as refracted through the media; *Basic Instinct* used the conventions of the neo-noir thriller complete with the mysterious and beautiful woman as a murder suspect, and what it is parodying is, arguably, psychological reductionism in the thriller. When we have witnessed the murder and identified the culprit through her actual body, what good is psychology except as a pointless puzzle? *Showgirls* begins with Nomi thumbing a ride and the man who offers her coming up with a few standard lines and indicating lascivious intent ("Come on! This is your lucky day," and leering). It ends the same way, the same man pulling up and uttering the same lines. The man robbed her earlier and the situation has altered; but my argument is that the film is deliberately serving up its world as cliché and faulting it on such grounds is simplistic. That being the case how is one to interpret a film that is deliberately constructed as cliché? Fassbinder used cliché deliberately and I read it as satire and perhaps the same is true of Verhoeven.

David Bordwell, as already indicated, differentiates between the kinds of meaning produced by a filmic text and suggests that the literal meaning is usually hinted at in a line of dialogue. In *The Wizard of Oz* (1939), it would be the key line "There's no place like home."[17] Such a line containing the conceptual meaning of the film would be uttered only by a sympathetic character not in a disturbed state since the meaning should be distorted neither by personal issues (e.g. a state of anxiety) nor by the hostility we might feel toward him/her. One such character is James Smith the African American dancer who, being a failure since he cannot use his talent fruitfully, also acts as a moral commentator. He uses strong but forthright language against the business interests in Las Vegas, comparing show business to prostitution. James is booed offstage during the song and dance routine at The Crave Club that he has banked upon, and marries a failed white dancer and she is also pregnant. Hollywood portrays African Americans in sexual relationships with white women but marriage is rarely mentioned[18]; more than marriage, the possibility of children from the marriage is taboo. Obviously, mixed marriages are not uncommon even in the industry but that is difficult in Hollywood's "unreality." This portrayal itself could be seen as indicative of the film's daring with regard to taboos.

The second such character is Zack, probably the only decent person Nomi encounters, also wryly clear-eyed about the business he is in. Zack is played by Kyle MacLachlan, the best-known actor in the film. Noni's only private moments are with him and she seems to genuinely care about Zack. At some point in the film, Zack tells her that he is an MBA. When Noni asks him what an MBA is, he replies, "It's a college degree one picks up that is worthless in the real world." A question one might ask here is what the "real world" is, whether Las Vegas is not part of it. Later in the film she is in his unreal-looking villa, both of them naked in a swimming pool, water gushing out of the mouths of stone dolphins. I propose that the film is making the point that "success" in the milieu involves immersion in a deliberate unreality and only failures like James Smith experience "reality." It is about a milieu that celebrates over-the-top behavior as a sign of attainment, which implies that Las Vegas is not the film's exclusive target. Millie, in Altman's *3 Women*, we recollect, tried to live a life meant for that unreal world but Altman approached that unreality through her social reality. Verhoeven is doing the opposite when he approaches the real world through the unreality of Las Vegas. "Unreality" is an un-theorized category but it has features in common with "postmodernity" *Showgirls* does not include the comical elements that abound in *RoboCop* and *Total Recall* and its satire has been missed but it is a portrayal of postmodern "unreality":

> One reason why we do not see the satire in *Showgirls* is that the film lacks the usual coordinates and signposts for a critique of human vice and folly provided by sarcasm, irony, and caustic wit. Such overt satire is safe because we "know" what is implied between the scare quotes and to whom it applies. Because this is not the case with *Showgirls*, critics often took the film as equivalent to or of a piece with its subject matter. According to (Anthony) Lane, if not most reviewers, *Showgirls* "has the distinction of being the first movie about Las Vegas that is actually more vulgar than Las Vegas."[19]

Calling the film "more vulgar than Las Vegas" obviously misses some of the points that I noted. *Showgirls* is more about postmodern America than Las Vegas and, to use Baudrillard's proposition about Disneyland as a simulacrum, it may exist to conceal the fact that it is the "real" country[20] or, rather, a "hyperreality." Baudrillard defined "hyperreality" as "the generation by models of a real without origin or reality"; hyperreality is a representation, a sign without an original referent; it is a copy for which there is no original and that could perhaps also be said of Las Vegas.

Starship Troopers (1997)

The film is based on the novel of the same name by Robert A Heinlen which was seen to act as a "vehicle" for the ideology of that era, evoking the paranoiac US politics of the 1950s in which, according to Senators Joe McCarthy and Barry Goldwater, America was "in clear and imminent danger of being overwhelmed by alien forces."[21] The film is set in the twenty-third century; while colonizing new planets, humans have encountered a hostile non-technological insectoid species known as Arachnids or "bugs" on the planet Klendathu. The bugs appear to be little more than savage, unrelenting killing machines although there are suggestions that they were provoked by the intrusion of humans into their habitat. The Arachnids live on a planet surrounded by an asteroid belt and they have the ability to force an asteroid out of an orbit and direct it in another direction. Verhoeven's film begins with a recruitment advertisement on the "Federal Network" for the Mobile Infantry, and in the United Citizen Federation, citizenship is earned by performing activities such as military service that grant individuals opportunities prohibited to basic civilians, like being in politics or having more babies. The protagonists are from Buenos Aires and after graduating from high school John "Johnny" Rico (Casper Van Dien), his girlfriend Carmen Ibanez (Denise Richards), and best friend Carl Jenkins (Neil Patrick Harris) enlist in the Federal Service despite Rico's parents' disapproval of military service. Carmen becomes a spaceship pilot, while Carl joins Military Intelligence. Rico enlists in the Mobile Infantry and is surprised to find Isabelle "Dizzy" Flores (Dina Meyer), his fellow ex-student, has deliberately transferred to his squad since she has a crush on him.

The actors who play the parts of the chief protagonists have facial features like those in romantic comic strips and graphic novels and that is evidently the tone Verhoeven is planning to maintain. The location of the story in Buenos Aires is also accompanied by its Americanization; the protagonists play American football rather than soccer, which one might expect in Buenos Aires. Verhoeven intersperses the story with announcements and notifications on the Federal Network—children made better citizens by being allowed to handle firearms, the announcement of a murderer's execution to be telecast on all channels. The sense conveyed is of persistent coercion through the media, especially of the young. In class, they are taught by a teacher with one arm, who they all admire, Jean Rasczak (Michael Ironside) who teaches them about citizenship. Before Rico enlists, he asks Rasczak's advice and is told to "make up his own mind," when, evidently, the persuasion in the media impedes independent thought.

The soldier who enlists him has a mechanical hand and assures Rico that the Mobile Infantry made him "the man he is." At the training camp where he is first placed, the trainer Sgt Zim is like Sgt Hartmann in Kubrick's *Full Metal Jacket*, although the trainees go along with his strictness. The grey uniforms worn by the trainees are reminiscent of German uniforms in World War II and Carl Jenkins, when he appears later as a colonel, is dressed in the attire of the SS! The overall tone is buoyant but there is a submerged sense of total war in which "brainwashed" people are crippled or are dying, and the juxtaposition is clearly satirical. If it has been read as an action film with a clear demarcation between "good" and "bad" that is Verhoeven's deceptive strategy, as we saw in his other American films beginning with *RoboCop*.

Rico and Carmen are both being trained in their respective fields, Carmen by their former classmate and Rico's rival Zander Barcalow (Patrick Muldoon). But believing that her relationship with Rico cannot be pursued, Carmen breaks off with Rico. Rico has been doing well as a trainee and gets promoted to squad leader but he makes a mistake which results in the death of a soldier and he is "administratively punished" with flogging. Disheartened, he resigns and intends to go back to Buenos Aires but the bugs send a meteor there and the city is wiped out along with his family. Rico therefore reenlists and an invasion of Klendathu is now underway to deal with the bug menace. There are now debates on the Federal Network and in the UN: "Humans and not insects must dominate the galaxy!" Children stamp out cockroaches on the streets as their contribution to the Human cause, evidently a nod to the xenophobia unleashed by war.

The invasion of Klendathu turns out to be a disaster; there are "missiles" coming from there, more lethal than intelligence made out. When they land, Rico and Dizzy discover them being launched from the behinds of giant bugs! Verhoeven is being viciously satirical here but he maintains such a straight face that no one can tell. *Starship Troopers* came out well before the Iraq War, but the idea of insects shitting explosive devices to bring down intergalactic spaceships anticipates what was alleged about Iraq, that it possessed Weapons of Mass Destruction that threatened the world. It is the familiar ploy of superpowers to allege that those they are planning to overrun represent a threat and this is essentially what Verhoeven is satirizing. Rico is injured; his death is even announced and the news duly reaches Carmen and Zander.

After recovering, he, Ace, and Dizzy are reassigned to the "Rasczak's Roughnecks," an elite unit commanded by Lt. Jean Rasczak, Rico's inspiring former high school teacher. He quickly gains the respect of his peers and is promoted to the rank of corporal after taking out a giant bug. The Roughnecks respond to a distress call from Planet "P," where they reconnoiter an outpost devastated by bugs, some of the dead soldiers having had their brains sucked out. The distress call is a trap, and the Arachnids swarm the outpost. Rico, now an acting sergeant, euthanizes a mortally wounded Rasczak after a buried bug bites off his legs. Dizzy is killed, but the surviving Roughnecks are rescued by Carmen and Zander. Rico and Carmen encounter Carl, now a high-ranking intelligence officer, at Dizzy's funeral. Carl reveals that there is reason to believe an intelligent "Brain Bug" is directing the other bugs and has been learning how to fight humans. He promotes Rico to lieutenant and gives him command of the Roughnecks, ordering the Mobile Infantry units under his control to return to "P" in an attempt to capture the Brain Bug. Carmen and Zander are captured by the bugs, deliberately kept alive for some purpose.

The climax of the film is the appearance of the Brain Bug preceded by various other kinds of bugs in the bug hierarchy. Verhoeven stages this as a ceremonial affair, the normally noisy and violent bugs becoming orderly in the presence of the Brain Bug as courtiers and officials do in the presence of the leader. The Brain Bug proceeds to insert its pincer-like proboscis into Zander's head and sucks out his brain: that is evidently the way knowledge of human thought is obtained by the insects! When it is about to do the same to Carmen, she cuts off its proboscis with a knife and it retreats, blowing air painfully. Rico arrives just then with a nuke and saves the day. The film concludes with the Brain Bug's capture and Carl trying to determine its thoughts: "It is afraid!" As an epilog is a Federal Network feature—the Brain Bug being "probed" to understand its thinking. What is being done to it is so ghastly that it is "censored." As already indicated, this film came much earlier but it anticipates Saddam and Gadaffi's horrible ends.

Where Heinlein delivers extended passages about the benefits of a social order organized by militaristic principles and does so without irony, Verhoeven subverts his book's conservative ideology by deconstructing military guts and glory even as it provides so much violence.[22] *Starship Troopers* is satire that has been read as being of fascism and Verhoeven has included Nazi uniforms as a design element. But much of it points to the film being a satire on American militarism, especially against technologically primitive enemies, and shows us how easy it is to generate violent patriotism. My own reading of it is that the successful war against the Arachnids refers to Iraq's destruction during the Gulf War of 1991. An array of nations joined the coalition, forming the largest military alliance since World War II. Most of the coalition's military forces were from the United States, with Saudi Arabia, the United Kingdom, and Egypt as leading contributors. The war marked the introduction of live news broadcasts from the front lines of the battle, principally by the US network CNN. The coalition forces could have been reconfigured as the "Federation" by Verhoeven and the Federal Network could be CNN.

Verhoeven being satirical in the films made in the United States with his audiences lapping them up—without recognizing the satire in them—has the effect that significant parts are not satirical but simply promote action. The satire and the action blended perfectly in *RoboCop* and, as I demonstrated, the nudity and sex in *Basic Instinct* had a justified purpose. Even *Showgirls* is more than justified as an artistic vision but not the second part of *Total Recall*, which is even banal when it deals with the oppression of the mutants by denying them air. Lavishing "humanistic" concern on an imagined category in fiction without an ulterior or allegorical purpose suggests an artistic block. *Starship Troopers*, too, may contain more violence than can be artistically justified.

Verhoeven's artistically most unsuccessful American film may be *Hollow Man* (2000), after which he went back to making films in Europe. This film, once again, is sci-fi, but it does not deal with systems as refracted through the media, being more straightforward. It is reminiscent of David Cronenberg's *The Fly* (1986) in that it is about a scientific experiment (here invisibility) gone wrong. But while the special effects are spectacular, there is virtually no satire in it. The protagonist, after making himself invisible, starts relishing the power it affords him. He works for the military but the film cannot be read, except half-heartedly, as about technology and political power. As one of his transgressions when invisible, the protagonist rapes an unnamed woman in the neighboring apartment but that woman plays no other part in the film and the segment seems inserted only as a bit of titillation. If the satirical impulse gets too clandestine and too few recognize it as such, there is the danger of it becoming extinct.

NOTES

1. James Reith, "*Turkish Delight*: Can the Sexually Explicit Dutch Classic Still Shock?" *The Guardian*, March 3, 2017. https://www.theguardian.com/books/booksblog/2017/mar/03/jan-wolkers-turkish-delight-a-uneasy-dutch-classic-of-sexual-candour. Accessed July 9, 2020.
2. Benjamin B Roberts, *Sex, Drugs and Rock and Roll in the Dutch Golden Age* (Amsterdam: Amsterdam University Press, 2017), pp. 12–13.
3. Roberts, p. 13.
4. Roberts, p. 14.
5. Roberts, pp. 15–16.
6. Roberts, pp. 27–29.
7. This was dealt with by Sigmund Freud in his 1920 book *Beyond the Pleasure Principle*. The life instincts are focused on the preservation of life, both of the individual and of

the species. This drive compels people to engage in actions that sustain their own lives such as looking after their health and safety. It also exerts itself through sexual drives, motivating people to create and nurture new life. Freud believed that people typically channel their death instincts outward. In support of his theory, Freud noted that people who experience a traumatic event would often re-enact that experience. From this, he concluded that people hold an unconscious desire to die but that the life instincts largely temper this wish.

8. "The petticoats and gown perfume, / Which waft a stink round every room. / Thus finishing his grand survey, / Disgusted Strephon stole away / Repeating in his amorous fits, / Oh! Celia, Celia, Celia shits!" Jonathan Swift, *The Lady's Dressing Room.*
9. In Greek mythology, Persephone is the daughter of Zeus and Demeter. She becomes the queen of the underworld through her abduction by Hades, the god of the underworld. The myth of her abduction represents her function as the embodiment of vegetation, which springs forth from the earth in spring and retreats into the ground after harvest; hence, she is also associated with spring as well as the fertility of vegetation.
10. Michael Robertson, "Property and Privatisation in *RoboCop*," *International Journal of Law in Context*, 4.3 (September 2008), 217–35.
11. Fred Glass, "The 'New Bad Future': *RoboCop* and 1980s Sci-Fi film," *Science as Culture*, 1.5 (1989), 35–36.
12. In *The Time Machine*, for instance, he saw humankind split up into two different categories, the Eloi and the Morlocks.
13. "Consider the last shot of the movie (no, I will not reveal it). This shot allows us to discover whodunit—whether one of the characters is a murderer, or not. The screen has faded to black. Then we get the last shot, and it answers our question. But if the last shot had provided the opposite answer, it still would have been consistent with everything that had happened in the film. Each and every shred of evidence throughout the entire movie supports two different conclusions." Roger Ebert, *Basic Instinct*, March 20, 1992. https://www.rogerebert.com/reviews/basic-instinct-1992. Accessed March 25, 2022.
14. Arlene Kramer Richards, EdD, "Woman as Medusa in *Basic Instinct*," *Psychoanalytic Inquiry: A Topical Journal for Mental Health Professionals*, 18.2 (2009), 270.
15. Alain J Cohen, "Stochastics of sex and death in Basic |filmic| Instinct," *Semiotica*, 112.1–2 (1996), 111–12.
16. Here is Roger Ebert who says exactly what I am saying that it is not erotic, but finds fault with it on that account: "'Showgirls' is the first big-budget 'adults only' movie in a few years and, to be sure, it contains so much nudity that the sexy parts are when the girls put on their clothes. It contains no true eroticism, however, and that's why I think it reflects a grounding in sexual fantasy: eroticism requires a mental connection between two people, while masturbation requires only the other person's image." Roger Ebert, "*Showgirls*," Roger Ebert website, September 22, 1995. https://www.rogerebert.com/reviews/showgirls-1995. Accessed July 19, 2020.

17. David Bordwell, *Inference and Rhetoric in the Interpretation of Cinema* (Cambridge, MA: Harvard University Press, 1989), pp. 8–9.
18. Stanley Kramer's *Guess Who's Coming to Dinner* (1967) is still considered radical for portraying such a marriage.
19. Chon Norega, "'A Whisper of Satire,' in 'Round Table: *Showgirls*'," *Film Quarterly*, 56.3 (Spring 2003), 36.
20. Here is the actual quote from Baudrillard: "Disneyland is there to conceal the fact that it is the 'real' country, all of 'real' America which is Disneyland [...]. Disneyland is presented as imaginary in order to make us believe that the rest is real, when in fact all of Los Angeles and the America surrounding it are no longer real, but of the order of the hyperreal and of simulation." Jean Baudrillard, "Simulacra and Simulations," in *Selected Writings*, ed. by Mark Poster (Stanford: Stanford University Press, 1988), p. 172.
21. Jamie King, "Bug Planet: The Frontier Myth of *Starship Troopers*," *Futures*, 30.10 (1998), 1018.
22. Philip Leonard, "Teach Phenomenology the Bomb: *Starship Troopers*, the Technologized Body, and Humanitarian Warfare," *European Journal of American Culture*, 25.1 (April 2005), 32.

6

Aki Kaurismäki: Irony and Hopelessness

Aki Kaurismäki is the filmmaker (among those included in this book) whose films come closest to empathizing with their protagonists—usually from the working class. Kaurismäki was born in 1957 in south-central Finland, the third of four children. His father worked for a variety of textile companies in seven different towns before Kaurismäki left home after completing secondary school. Kaurismäki left his compulsory military service in Finland's army incomplete and applied to Finland's film school in 1977. The school refused him admission and he studied journalism at the University of Tampere but did not complete the degree.

Between 1978 and 1981, his older brother Mika studied filmmaking at Munich's University of Film and Television. Mika's final project furnished Aki with an entry into filmmaking. Aki co-authored the screenplay for *The Liar*, in which he also played the protagonist. The film won the Risto Jarva Prize at Tampere's short film festival in 1981. Although the film sold only some 500 tickets in its theatrical run, it made an impact on the intellectual class. *The Liar* signaled the Kaurismäki brothers' distinctive approach to cinema. They leveraged the prize money and prestige of their prize to finance three further projects between 1981 and 1982.[1] Aki Kaurismäki made his directorial debut in 1983 with an adaptation of Dostoyevsky's *Crime and Punishment.* Given the paucity of available public knowledge about Finland, its culture and politics to most film students, the best way to get a purchase on Kaurismäki's approach and concerns would be through this film and another film adaptation of *Hamlet,* since they are not adaptations as much as subversions or parodies, especially *Hamlet.* Kaurismäki's comedies and his working-class dramas can then be studied as separate categories. It is the last category that has won him the greatest renown in world cinema.

"Adaptation" of literary classics

Crime and Punishment (1983)

In the film, Antti Rahikainen (Makku Toika) is a slaughterhouse worker and the film begins with him cleaning up at the end of the day. He is then shown waiting in the shadows for an upper-class executive Honkanen to arrive in his car and go up to his apartment. Antti follows him upstairs, rings his bell, and pretends to be there with a telegram. When the man goes inside to fetch a pen to sign the acknowledgment, Antti follows him inside and shoots him dead. We are not told why but he stays on after, taking the victim's watch and wallet. Some moments later, a young woman Eeva Lakso (Aino Seppo) enters the house and finds sitting him there. She is there to cater a party and Antti admits that he killed the man; she asks him to leave. Antti leaves later and deposits whatever he has taken from the man in the locker at the railway station.

Antti is a lonely person and stays away from work thereafter; his only friend is a co-worker played by Matti Pellonpää, later a Kaurismäki favorite. Eeva is questioned by the police but she reveals nothing. But rather than keep away from her, Antti goes to meet her and grows friendly. Eeva's employer has been courting her but she is drawn more to Antti. After some investigation, the police come upon the knowledge that the murdered man had killed someone in a drunk driving hit-and-run case but was acquitted for want of evidence. Since Antti was the fiancé of the victim and had threatened revenge, he has a motive. Antti now decides to leave Finland and gets a forged passport. His friend in the slaughterhouse played by Pellonpää is learning English and has plans of leaving too. As part of his plan, Antti frames a beggar in a street corner by giving him the key to the locker with the murdered man's belongings and having him apprehended. The suspect is so weak mentally that he even confesses to the murder that he is innocent of. Eeva matters to Antti but after sheltering him from the police and knowing for certain that he is Honkanen's murderer she is dismayed that he framed an innocent man. The film concludes with Antti deciding not to leave Finland but taking responsibility for the murder; he is sentenced to eight years. Eeva says she will wait for him but that is a long time and he dissuades her. "I killed a louse but became one myself," he admits.

Crime and Punishment has not been widely written about but a conspicuous aspect in this first film is Kaurismäki's use of music. Parts of it are diegetic, played often in cafés where Antti sits alone or with Eeva. The music is usually American pop or jazz and is romantic, misaligned with the mood of the scene. The effect is difficult to describe but it conveys the sense of an artificially created ambience that has little to do with the lives of those inhabiting the space. Then there is also occasional operatic music heard on the soundtrack at heightened moments. This again is unsuitable since these are very ordinary working-class lives and the music is too grand.[2] Antti actually attends an opera and fidgets through it to emphasize the point.

My own reading of the use of music is that Kaurismäki is ironically distinguishing his own vision from Dostoyevsky's. Where Raskolnikov was highly articulate and a philosopher (of sorts) in Dostoyevsky's novel, Antti and Eeva are inarticulate and mostly silent. Kaurismäki's long silences are not pregnant with unspoken emotions but are the only evidence of his characters having very little to say. My proposition is that Kaurismäki is, in a sense, parodying Dostoyevsky's novel rather than "adapting" it.[3] When first intimated by the police inspector of the possibility of his hand in the murder, Antti pretends that it was done with an axe, when he actually used a pistol. Antti works in the slaughterhouse and sees blood flowing abundantly every day. Raskolnikov (who used an axe) was genteel (though poor) and the sense to be got is that an axe would not be the right weapon for a slaughterhouse worker; murder, being extraordinary for him, would necessitate a more unfamiliar weapon. Dostoyevsky arguably made Raskolnikov use an axe because it would be furthest from his gentility.

Lastly, there is palpable dissatisfaction in Kaurismäki's feelings for his home country though the reasons are not articulated. Leaving it is a primary impulse for both men and Eeva will accompany them. Kaurismäki is celebrated in Finland but that apparently angers him,[4] and his feelings show up. Most of these aspects of *Crime and Punishment* recur in his mature films as part of his filmmaking style.

Hamlet Goes Business (1987)

Kaurismäki's adaptation of Shakespeare is more faithful to the play though it is also parodic in effect. In the film, Hamlet is the scion of a business family and when the film begins his father is poisoned to death by Klaus, presumably his brother, who is also in a clandestine relationship with Gertrude, Hamlet's mother. With his father's death, Hamlet inherits 51 percent of the shares but Klaus (now the president) does not consider him a threat, since Hamlet has been a wastrel, spending his time trying to coerce Ofelia (Katti Outinen) into sex. Ofelia is being prepared to become Hamlet's wife by her father Polonius (Esko Nikkari) and instructed not to yield to Hamlet. Klaus, meanwhile, is trying to push for a business deal with a Norwegian corporate to give up his shipyards and sawmills and primarily concentrate on making rubber ducks.

The "rubber duck" turn is evidently parodic and Kaurismäki makes it evident from the start that he has no sympathy for any of those from Hamlet's family or business, including the dissolute Hamlet himself. When his father's ghost appears, it lets Hamlet know that it thinks lowly of him but hopes that he still loves him enough to take revenge on Klaus for his murder. The film follows Shakespeare's play closely enough but, unlike *Crime and Punishment*, it is wordy—as Shakespeare was. At the same time, the words do not lead to useful communication. When Hamlet acts mad[5]—attributed by Klaus to sexual frustration at being denied Ofelia—the nonsense he spews is not distinguishable from his normal communication.

Hamlet is cunning enough to catch on to Klaus's machinations but hardly a psychologically ambiguous character—like Shakespeare's. His isolation/alienation within the family is shown when he is with his mother at the ends of a long dining table even longer than the famous one in *Citizen Kane* in which Kane and his estranged wife have dinner, and which can only be read as parody.[6] But a character Kaurismäki is in sympathy with is the working-class Simo (Hannu Valtonen),[7] Hamlet's long-suffering driver and the equivalent of Barnardo in the play. A key motif in the film is the likely effect on the labor of closing down the shipyards and Simo is entrusted with keeping an eye on Hamlet by the union. When Hamlet, after vanquishing all his foes and seeing them dead, takes charge of the business and reiterates his intention to sell it away, most likely resulting in the laying off of poor workers, Simo kills him.

Kaurismäki's use of music is also ironic here as in *Crime and Punishment*. When he is with Ofelia, he plays some classical music—that might be suitable for someone cultured—but turns off the tape and kicks a jukebox into raucous pop music when she leaves. Hamlet meets Rosencrantz and Gyldenstern in a pub later and Finnish rock band Melrose is playing their song "Rich little Bitch."[8] The gyrations of the group are comic but Kaurismäki is using it to suggest that it is still more immediate than the high culture represented by Shakespeare's complexities. I infer this from the pointless relationship between Hamlet and his two "best friends," essentially corporate lackeys being used in boardroom machinations. Rosencrantz and Gyldenstern also look alike to imply that people who play such a role are not distinguishable.

Hamlet Goes Business shows control over film craft but suggests a filmmaker not interested in his subjects, except in the consequences of their machinations. There are many deaths in the film but the police, though mentioned, never appear, denoting what money can buy. Norway, Britain, and Sweden are mentioned, but the primary implication of international business is that it is unholy.[9] The ways of corporate houses and big business, Kaurismäki lets us know, do not particularly involve him. But unfortunately for him, the working class, with which he is in sympathy,[10] has hardly a presence in Shakespeare's play and when he tries to include it through Simo and the Guard (Matti Pellonpää), in whose company Hamlet sights his father's ghost, there is still too little. Kaurismäki is coercing Shakespeare's text to say something very different but one does not see much of a fit between his concerns and the vehicle he employs.

Comedies

Aki Kaurismäki's deadpan style of filmmaking is suited to a brand of comedy related in some way to Buster Keaton. The difference is that in Keaton the deadpan was an attribute of the character while here it is a quality of the filmmaking. The broadest of his comedies may be *Leningrad Cowboys Go America* (1989), which was also a success. The Leningrad Cowboys also feature in some later comedies but I will consider only this first one, perhaps still the best.

Leningrad Cowboys Go America (1989)

The film, by all accounts, was inspired by *The Blues Brothers* (1980) and *This Is Spinal Tap* (1984). The protagonists in the film are a band from Siberia and it is not immediately clear why Kaurismäki moved away from Finland for his protagonists. As indicated, he has working-class sympathies that may have seen him drawn to the USSR. But in 1989 (when the film was made), the Eastern Bloc was in a state of collapse and the USSR was in desperate economic trouble. Having a Soviet rock band (with such a name), it has been suggested, was a joke on the USSR.

When the film begins, the group is busy performing for an agent who pronounces their work "shit" and suggests they go to America, where "they will take anything." He also provides them with a contact in Manhattan. The members are possibly farmhands—since they are seen driving tractors—which suggests that they were working on a collective farm that is closing down.[11] The establishment of the group could then be seen as an ironic effect of Gorbachev's Perestroika. One of the members has frozen to death during the rehearsal and his father, while helping the group out with money, directs them to take him along with them to "see the world." His coffin therefore keeps them company on their long journey across the United States.

To describe the group, all of them including their manager Vladimir (Matti Pelonpää) wear their hair protruding ahead of them and ending pointedly upward. Their pointed shoes also turn sharply upward. Everyone associated with them—including the frozen member's father and his dog—sports the same hairstyle. Igor, the village idiot who wants to join them, is kept out because his hair is short. Their frozen comrade's coffin has the samepointed hair, the shoes, and an electric guitar protruding. The members are capable singers but their efforts at giving themselves individuality are eccentric; their failure is not due to incompetence but to their outmoded music.

Leningrad Cowboys is a road movie and their journey commences on the airplane with all of them learning English, the language of the film. Igor is a stowaway. In Manhattan, their potential agent says that their kind of music is only good in Mexico, directs them to a cousin's wedding there. The group realizes that they must learn rock and roll to succeed and they buy an old Cadillac to travel southward, their frozen comrade in a coffin packed with ice—in which Vladimir keeps beer that only he knows about. Much of their journey is through the Deep South where they play at sad places with a poor clientele, mostly black. Igor follows them and is allowed into the group only in Mexico. Igor also unfreezes their comrade and brings him back to life. Vladimir, who has been exploitative, leaves them and disappears but their music is a smash hit.

On the surface, there is not much irony in evidence in the film but America is not presented as glamorous despite it initially offering "hope" to the group. The towns they travel through are not prosperous and the places where they play are full of glum people, conspicuous when compared to the good cheer palpable in Mexico. The celebratory music the group plays is hence at odds with the mood. But there is another aspect that has to do with the genre of the road movie which is basically about people trying to "discover a country" by traveling through it, but usually being pessimistic. Peter Fonda's *Easy Rider* (1969) and Ridley Scott's *Thelma and Louise* (1991) are American road movies and the protagonists are killed at the conclusion. *Leningrad Cowboys* begins with refugees from the USSR (the film does not describe them as "refugees") coming to the "promised land," discovering it by traveling through it but leaving it eventually to find the "promised land" only in Mexico. That, I would argue, is not only comedy but also irony, since America is where they were directed. The fall of the Eastern Bloc was a conclusive victory for the West and Kaurismäki presents a picture of the United States at odds with the triumph.

Calamari Union (1985)

Kaurismäki made an association between working-class people and the musicians who entertain them in *Leningrad Cowboys Go America* and the same association is made in *Calamari Union*, described as an "absurdist" or "surrealist" comedy. The musicians in the film are especially punk in characteristic. Punk in Finland was an attack on institutions of the middle class but in its early form it came from within the middle class, in contrast to the movement's working-class constituency in England.[12] *Calamari Union* begins late at night in a café where a man is drumming away until summoned to join others sitting at tables. It would seem that the seventeen men present—who are all called Frank—are members of the band that has been playing there. One of them—presumably to assert his individuality—is called Pekka and speaks English, presumably helping himself to dialogues from American films since a line he repeats is Travis Bickle's "You talking to me?" from Martin Scorsese's *Taxi Driver* (1976). Their leader, reading from a paper, declares that their district in Helsinki (named Kallio) is too oppressive for a variety of reasons including the poverty and the pollution in the air, and he sets them the task of making across the city to Eira district by the sea, ostensibly still unspoilt.

It is difficult to keep track of the chronology but the men set out together stealing a metro train for part of the journey but gradually lose their way, are killed (shot, drowned, or by suicide), or distracted by other things. One steals a bag and gets into a hotel room, two more doze off in a cinema while others find themselves in less glamorous locations. One falls willingly into a river and drowns, prompting his companion to leap onto the bonnet of a passing car, which spends the rest of the night driving around the city before dumping its passenger in front of the café that the group has selected as a meeting place. Midway through the film, those remaining perform as a band but break up thereafter. Eventually, only two of the members arrive at their destination, by their own admission several years too late since it is no longer pristine. After a tussle both climb into a rowboat and commence rowing toward Estonia.

The film has been dubbed surreal and compared to Buñuel's work[13] or absurd by others but it is also interpreted as being about working-class solidarity:

> it is clear that "the working class has no fatherland," as one of the characters in the film suggests [...] The abstract adventure film *Calamari Union* (1985) functions as an illustration of the ways these films challenge the mental and social ecologies of neo-liberal capitalism. The film is at its base a capitalist critique of the organization of the city space into zones of deprivation and affluence. It focuses on a group of identically named men, the Franks, who awaken out of the slumber of their subjugated, oppressed existence to seek a way out of the geopolitically coordinated confines of the city.[14]

But while admitting the above, the interpretation must also take into account the fact that the numerous Franks, while setting themselves a simple task which is to reach a distant part of the city, are unable to do so without being distracted, which implies a satirical take on the subject. "Calamari union" implies a union of squids although what aspect of "squid solidarity" it pertains to is uncertain. Finally, as in many of his other films, hope is also to be found outside Finland, here in Estonia which was then part of the USSR.

Working class dramas

None of the films dealt with hitherto suggest a great filmmaker and they are perhaps too whimsical to turn the filmmaker's sensibility into a valid method but

Kaurismäki comes into his own with his working-class dramas. His working-class dramas can be profoundly heart-breaking even when they conclude optimistically and it may not be an easy task locating their irony. I am going to look at only three films in chronological order including the film I take to be his greatest, *The Match Factory Girl* (1990). Kaurismäki achieved his biggest international success later with *The Man without a Past* (2002), a film I am not examining since his later films do not pack the same ironic edge. I will however touch upon a slightly earlier film from the optimistic phase *Drifting Clouds* (1996) to understand the difference and speculate about it.

Ariel (1988)

Ariel begins with a mine dynamited as a prelude to its closure. Taistu Kasurinen leaves his workplace for the final time with an older colleague. After a drink, the colleague curses the life around him and hands over the keys of his car to Taisto as a gift. He warns Taisto that the step he proposes to take is no better, cocks a pistol, disappears into the cloakroom, and Taisto hears the gunshot while he smokes. The sequence does not induce the standard emotions aroused by a working person's suicide but neither is it unsympathetic. It is matter of fact—as if the hopelessness of the situation would render pathos incongruous. Taisto now has the dead colleague's Cadillac, which is a convertible. It is a handsome legacy but the top will not go up, an inconvenience in winter.

The episodes provide clues as to why Kaurismäki's is essentially a comic and ironic vision regardless of its inherent pathos. Where the situation described would normally lead to pity being the dominant emotion, he inserts incongruous bits to direct us away. This is not "optimism" or "seeing the bright side" but implies that pathos being an emotion appropriate for narrative climaxes, it should not be evoked when dealing with lives that are customarily desperate. Where *Bicycle Thieves* moves from hope to despondency, *Ariel* begins in desperation. It is one of the conventions of realism that misery should invite empathy and Kaurismäki breaks it perhaps because "empathy" does not help those he makes films about; the dead man's situation was so hopeless that he took an ironic distance from it as if it were someone else's.

In *Ariel*, Taisto draws out all his savings and goes to the city. When he takes out his wallet to buy himself a hamburger, two men catch sight of his money. As he munches on his purchase, the two men compliment him on the car and want to look at the engine. When Taisto obligingly opens the hood, one of the men casually swings a beer bottle at his head and Taisto is left without his savings. But all is not lost because Taisto finds work on the docks the next day. There is even a fur coat to keep him warm because its owner just fell under a forklift, an ironic commentary on the usefulness of another worker's misfortune to oneself. Taisto also finds accommodation in a flophouse. His neighbor has put up an icon of Jesus on Taisto's part of the wall and Taisto responds by putting up a picture of a nameless personage—stolen from a photo studio—to mark his own territory; that in any case was the purport of his neighbor's icon.

Ariel is a love story involving Taisto and Irmeli, whom he meets when she hands him a parking ticket and they end up having dinner together. Irmeli has been deserted by her husband but she has a son and works at several jobs including handing parking tickets and in a slaughterhouse. Taisto soon encounters the man who stole his savings but goes to prison for attacking him and trying to retrieve his money. In prison, he decides to marry Irmeli and has a brass wedding ring made for her in the workshop by Mikonnen (Matti Pellonpää), there for manslaughter and sharing his cell.

Breaking out of prison is easy but there need to be plans for after that—or one will get only to the nearest tavern; going southward to Africa or South America seems attractive. But this means getting fake passports or being stowaways on a ship. If the goods start moving about in a storm, a stowaway in the hold could be killed. Taisto with Mikonnen does all this though it involves robbing a bank along the way. Mikonnen is however killed in the process when the middleman tries to cheat them. As he sits dying at the wheel of Taisto's open car, he instructs Taisto that when he is dead his heart should be buried in the garbage dump. But before breathing his last, he also accidentally discovers the button that makes the convertible's roof go up. The film ends with Taisto, Irmeli and their son becoming stowaways on the *Ariel*, bound for Mexico.

My effort in the above description of *Ariel* is to convey a sense of the film's laconic appeal. The film is optimistic but it is a strange "optimism" because it implies the impossibility of being in the home country and also hopeful. Prison is not brutal by the standards of prison dramas but it saturates each inmate with hopelessness. Taisto's birthday cake is mangled beyond recognition by the guards looking inside it for implements. The film makes these observations with deadpan bitterness, and its attitude can be seen as blank-faced rage. Films like *Bicycle Thieves*, by and large, exhibit an attitude of powerlessness toward social authority and *Ariel* seems similar since Taisto submits to authority without demur. But things work out for him like the successful bank robbery, his shooting dead the middlemen after they have knifed Mikonnen. Taisto is also able to leave for Mexico (with money) with Irmeli and her son. Irmeli's son is a touching invention since he is so obliging and uncomplicated, preoccupied with his comics.

Given the conditions that the poor labor under, Kaurismäki is thoughtfully putting providence on Taisto's side. There is an unwritten convention that people who have perpetrated crimes should face punishment/retribution and, in *Bicycle*

Thieves, Ricci could not have escaped with the bicycle. In caper films, criminals are allowed to get away only if they have robbed someone not entitled to the loot as in *The Italian Job* (2003). This is largely a characteristic of films from deemed democracies[15] where generic conventions have incorporated it, and I would see it as an expression of faith in the fairness of the state. Kaurismäki is from a democracy but in film after film he shows little faith in his home country—but without any utopian suggestions that things might have been better under a different political dispensation.

This leaves us with the perception that *Ariel* is an exercise in humanist realism in which the passive protagonists accept their lot although the social conditions are entirely unfavorable—and with no hope around the corner. Still, while Taisto submits initially, he also shows what he can do if required to.[16] Much of what he undertakes when he learns to "cope" would not be permissible in humanist cinema—examples would be from the Italian neorealists—since they are criminal acts. Using his minimalistic approach, Kaurismäki makes robbing the bank seem an uncomplicated exercise although Mikonnen is killed by the criminal middlemen who try to cheat them. Where providence is shown to work against the criminals in most films, their plans bear fruit smoothly for Taisto. But the factor of importance is that at the conclusion, Taisto, Irmeli, and the boy are en route to Mexico and with the money to see them through.

Ariel has been included as part of a "loser trilogy" (thus named by Kaurismäki)[17] but is Taisto a "loser"? Kaurismäki actually exults in Taisto's escape—not only implying that he has no faith in the system but also in the possibility that a better system might replace it. In his attitude to social structure, he emerges as an anarchist like Buñuel; but rather than merely critique it, he is concerned with how people should cope with a system that they have no chance of righting. But since *Ariel* has such a heart-breaking tale, where does its irony lie? It arguably lies in Taisto being a "loser" when he has faith in the system and functions by an ethics acceptable in a democratic society but triumphs when he breaks the law and escapes it. In any "humanist" film other than from Kaurismäki, Taisto would probably have to face retribution in Mexico![18]

The Match Factory Girl (1990)

The Match Factory Girl (1990) is arguably Kaurismäki's darkest "comedy"; it is a comedy since he takes a greater distance from his woman protagonist than he normally does and sees her faulty judgments. The film begins with a quote from a trashy romantic novel[19] and, in this film, Liris (Kati Outinen), who lives with her sullen mother, and ill-tempered step-stepfather, works in a match factory, spends her free time reading romantic novels and her weekends dressing up and waiting in a café for the right man. The first sequence shows pine sheets cut up to produce matches and Liris is as nondescript as a matchbox, it seems. Kati Outinen has a physiognomy comparable to Shelley Duvall's—as Millie in Altman's *3 Women*—but Millie was gregarious, as Liris is not. Altman's film was the tragicomedy of a gregarious person avoided by the others, but Liris is different here. The money Liris earns is taken away by her father and she is slapped for spending some of it on a dress. A telling scene at a dance hall shows her sitting alone in her new dress with empty soft drink bottles beside her. She has a brother who has left home because he cannot get on with his father and he works in a restaurant as a cook.

The man Liris eventually meets is Aarne, a yuppie, who mistakes her for a prostitute. The next morning we see him dressing up in a business suit and we catch on to the class differences between them.[20] Not seeing the money left for her, Liris goes home with a warm smile after leaving a note for him with her telephone number; but Aarne does not call. She visits him and he pushes her away, promising to call the next evening. The sequence in which Aarne calls on Liris is Kaurismäki at his most savagely ironical. The step-father is dressed ceremonially in a suit and tie though still without conversation; suit and tie to Aarne are work attire. The family evidently believes that Aarne is "Liris's man" but he is taking her out only to clear up matters. Aarne sizing up her family's station by glancing around tells us how the gap between the two is unbridgeable.[21] When Liris finds she is pregnant, Aarne will still have none of her. Her hopeful suggestion that raising the child together "will bring love" does not tempt him. Liris has just written a long letter to him explaining the joys of parenthood and carried it to him herself—"to prevent it from falling into the wrong hands," but Aarne simply offers her more money "to get rid of the brat," and even that curt message is sent by post. Liris's mother reads the note—that Liris has been too distraught to hide—and she is asked to move out and be on her own even as she is in hospital after a traffic accident—an attempt to kill herself.

Kaurismäki's sympathies are obviously with Liris but he takes her viewpoint less than he took Taisto's in *Ariel*. He describes her desperate situation with lucid distance so that we understand how mistaken her judgments are. We know from their first moment together that Aarne is indifferent to her and, when he visits her in her home it is not as a suitor. The gap between Liris's real situation and the way she imagines it is so large that her situation is "comic" although it will end badly. When Liris finally understands that Aarne has no intention of accepting her child, her response is vintage Kaurismäki. She makes for Aarne's apartment to say good-bye and sweetly laces his wine with rat poison (she earlier bought a "large pack"). She then proceeds to a café where another man tries to be "friendly" and amiably pours rat poison into his drink. She now returns home and feeds rat poison to her father and mother. The film ends in the match factory with Liris being led away by the police. As might be anticipated, there is romantic music, this time about love and its disappointments, on the soundtrack.

I described Kaurismäki's emotions in *Ariel*—underneath his minimalistic style—as "blank-faced rage" and he transfers the rage to Liris in the film.[22] That her rage is not sharply directed is evident, but I propose that the same is true of Kaurismäki in that he has no diagnosis to offer for the ills he finds in his society, which is perhaps what brings him closest to Buñuel—as in *Las Hurdes*. Earlier I described the use of popular music in his films as ironic but when we come to *The Match Factory Girl*, the banality of the romantic lyrics becomes "blank-faced" like Kaurismäki's rage. They mean nothing in themselves but are a gesture toward what cannot be expressed just as Kaurismäki's minimalism can be interpreted as unspoken rage.

This last assertion associating his minimalist style with "unspoken rage" may be questioned since his earliest films use the same style but without anger. Still, I propose that in *The Match Factory Girl*, Kaurismäki uses diversionary music or keeps the events off-screen pointedly when they signify the protagonist's hopelessness or rage—like Liris's attempted suicide; she lacing people's drinks with rat poison but the consequences happening off-screen; she playing a jukebox in her brother's apartment when cast out by her parents with nowhere else to go; and her warm smile thinking of Aarne's love when we know his true feelings better. Kaurismäki's use of silence in conversations is reminiscent of the early Fassbinder (*Katzelmacher*)—rather than Bresson's *Mouchette,* which he admitted drawing from. Its import is not that words are inadequate but that people have little to communicate, although Kaurismäki is more sympathetic to Liris than Fassbinder was to Jorgos. I would also propose that a silent "exchange" in *The Match Factory Girl* signifies two people at odds but not knowing it. It would take more substantiation than is possible here but it could be argued that his minimalist "short-cuts" are much more used in the service of anarchic emotions in this film than earlier.

The two films dealt with in this section hitherto may leave the reader still wondering about the "irony" in Kaurismäki's films. While that was obviously true of *Hamlet Goes Business* and the director demonstrated no empathy in that film as he does in these two, one might wonder why these two are not simply termed black comedies; but there is more here than opportunity for incongruous laughter. Unlike the chapters on Buñuel, Fassbinder, or Kubrick, there is less "interpretation" in whatever has just been said about them and one cannot mistake what Kaurismäki means. My own sense of them depends on the fact that Kaurismäki not only comes from a democratic milieu but his films also admit that openly. Taisto going to jail in *Ariel* for trying to recover his money from a thief is "just" since the law only proceeds on the basis of available evidence and there was none obtainable in his support. In *The Match Factory Girl*, Liris, although we are in sympathy with her, is arrested *rightly* for murder. In neither of the films is a miscarriage of justice lamented by Kaurismäki. A clue was found in the happy ending in *Ariel*, providence being on Taisto and Irmeli's side. In *The Match Factory Girl*, a similar "happy ending" would evidently not have been plausible, given the seriousness of Liris's crimes. Kaurismäki's rage in these two films is actually against the "natural order" rather than against a politically correctable system and he is aligning providence against it in *Ariel* since there is no other recourse. The speechlessness of the characters—and their inability to articulate solutions for themselves—is then a measure of their helplessness against this order that they cannot even protest since it is one arrived at entirely *by democratic consensus.*

Kaurismäki's anger—unlike Altman's or Buñuel's—is less clearly directed against any identifiable characteristic of his society but it is greater anger than his wronged protagonists feel. *The Match Factory Girl* is ironic because Liris has so much more faith in it than Kaurismäki has—given her expectations from Aarne who belongs to a different class and the gap between them unbridgeable. Her murderous rage then becomes a token of this late realization. We could say that if Kaurismäki empathizes with his protagonists, it is still tinged with irony at their misplaced faith in the social order. In instances of empathetic cinema, this faith in the social system is not problematized, as for instance there are no faulty judgments on Ricci's part in *Bicycle Thieves.*

Take Care of Your Scarf Tatiana (1994)

This is perhaps Kaurismäki's happiest (and funniest) film though it is more an awkward love story than a comedy. Valto (Mato Valtonen) is a tailor who lives with his mother in a small Finnish town. When he runs out of coffee—to which he is addicted—and the domineering woman will not replenish it, he sets off to get himself some more—after duly locking her up—and also a coffee maker for his car, which is being repaired by Reino (Matti Pellonpää). The two then set off on a long "test drive" lasting several days and nights. In a café frequented by foreign tourists speaking English, the men are accosted by two foreign women, Tatiana (Kati Outinen) from Estonia and Klavdia (Kirsi Tykkyläinen) from Alma Atta in Russia. The women want a ride to the port to catch the ferry back to Tallinn (in Estonia) and only Tatiana speaks Finnish.

Tatiana is essentially a road movie about the foursome, the men not communicating with the women but drinking coffee and alcohol. Each man shares a room with a woman in a rural hotel—in which the owner is also a receptionist and maid—but they never exchange words. Reino dislikes men from Helsinki—who act like "rockers"—and picks up a fight, but nothing much results. Reino is the more talkative but one rarely catches the gist of his anecdotes. The men make jokes between themselves—Reino going off into suppressed giggling—or discuss car-repair tools, but they say little to the women, who converse in Russian. In the evening at the hotel, the women dance together while the men sit silently over coffee and vodka. Still, Reino and Tatiana come together and, after going over to Estonia, Reino stays on. Valto returns home with another coffee maker and back to tailoring, freeing his vexed mother before commencing to stitch.

In being a rural film rather than one set in Helsinki, the film also changes its approach from the films discussed earlier. Kaurismäki seems to share Reino's antipathy toward Helsinki:

> Kaurismäki comments: "Starting in the 1960s, under the influence of developers, inept politicians and economists have almost entirely destroyed the Finland that I had become attached to—agrarian Finland, but also the Finland of Tatiana, the architecture of its towns and villages. The country has been transformed in a kind of globalisation which has left neither a roof nor a head untouched."[23]

This takes us to a new possibility we have not discussed in his films which is the domicile of his protagonists. Taisto in *Ariel*, we saw, is a migrant into Helsinki and spends his nights in a flophouse before meeting Irmeli. In *Man without a Past* (2002) also, the man is from outside, waylaid by hoodlums in Helsinki and caused to lose his memory. But if we look at the quote given above, there is nostalgia for Finland, as it once was, in most of his films. Taking this into account, the differences between the country and the transformed city could then correspond to that between the remembered past and transformed present of today; and Kaurismäki crowds his films with nostalgic references to various cultural artifacts including music and fashion:

> Kaurismäki's nostalgia is a means of creating an oppositional position exterior to economic and political discourse and institutions. In this view, the power of Kaurismäki's intervention lies in the position's temporal exteriority. Aligned with the forgotten past, Kaurismäki can supposedly see differently than the empowered of the present day. He does not depend on social networks or the rules of the system, and so can speak freely. And he is liberated from conflicts of interest, which might silence him. Nostalgia hence serves to vouch for the moral soundness of Kaurismäki's critique.[24]

Traces of this nostalgia can be found in most of the films dealt with and in *Calamari Union*, for instance, the group bemoans what has happened to their city and wish to traverse it to reach a seaside area, still ostensibly unspoilt. Only in *Hamlet Goes Business* do his protagonists fit neatly into present-day Helsinki, but that is also the only film examined in which Kaurismäki has no sympathy for them. But in the light of this factor—his work being regarded as suffused with nostalgia—how does one interpret *Take Care of Your Scarf Tatiana*, viz. its road movie design?

While dealing with *Leningrad Cowboys Goes America*, I noted some aspects of the road movie as a genre—that it is essentially an exploration of national space, and that there was irony in the group seeking to conquer "America" but finding success only in Mexico. *Take Care of Your Scarf Tatiana* was made in 1994 and has more aspects in common with *Leningrad Cowboys* than generally noted. Most conspicuously, through Tatiana and Clavdia, it also invokes the Soviet Union. There is a Soviet post office in Tatianna's village where the three arrive after dropping Clavdia off at the train station. The women are both Soviet citizens and the solidarity between them is perhaps of the "terminal" kind since Estonia became independent shortly after the film was made. Reino and Valto were to have dropped off the women at the port but they cross over with them to Soviet Estonia, and Reino stays on with Tatiana. She is living poorly by most standards since we see three boys standing at the door of her building; Tatiana is sharing a rather small space with other families. Still, Reino chooses that over returning home with Valto. Considered alongside *Leningrad Cowboys*, one could say that Kaurismäki is regarding the USSR also with nostalgia. Both road movies see the end of the Soviet Union with melancholy. Given this fact, one wonders if there was an unarticulated faith in a Marxist utopia in Kaurismäki's earlier films that he (wisely) remained silent about.

Later optimism

On the surface, Kaurismäki continues in the same vein with working-class subjects in his later films but one still senses a change; he seems to become optimistic. Each of the three films dealt with hitherto in the last section began starkly in the workplace. *Drifting Clouds* (1996) also begins in the workplace but this time it is an elegant restaurant and an African American singer is playing on the piano, singing a love song in English. This is different from the match factory as worksite or Valto bending over his sewing machine in *Take Care of Your Scarf Tatiana*. In *Drifting Clouds*, Ilona (Kati Outinen) is the head waiter at the "Dubrovnik" and her husband Lauri is a tram driver. The film begins with a crisis at the restaurant since the cook Lajunen is drunk and brandishing a knife. Malartin, Ilona's assistant, tries to get close but has his wrist cut up. But Lajunen is disarmed and order restored. A short while later, the "Dubrovnik" is taken over by a chain and the genteel owner Mrs Sjoholm reveals that the chain bought her out through the machinations of a bank manager and it cannot use the present employees.

Unlike the earlier films—with their harsh aesthetic—*Drifting Clouds* is elegantly shot in the most charming parts of Helsinki. It is affecting as a story of poor people struggling against odds and one searches for Kaurismäki's customary irony. The earlier owner of the Dubrovnik bankrolls a new enterprise for its former employees and it succeeds. I noted earlier the absence of utopian alternative in his films and the touch of naïveté that most political cinema exhibits. Kaurismäki believes in working-class solidarity but there were nonetheless admissions of irresolvable antagonisms. As instances, Reino giggles rudely when Tatiana and Clavdia approach the men in *Take Care of Your Scarf Tatiana*, justifying Clavdia's labeling them "dumb Finns." Clavdia even hints that her wildly drunk former husband Volodya was better company than Valto, with whom she shares a room. In *Ariel*, Taisto puts up a picture on the flophouse wall demarcating his territory from his neighbor's. *Drifting Clouds* has the drunken Lajunen brandishing a knife at his colleagues but he sheepishly compensates Malartin for his injury later. The scene is touching but one wonders if someone violently addicted to alcohol would be a good associate in a new business venture.[25] It is as though Kaurismäki was wishing away discord between his "loser" characters in *Drifting Clouds*. But most importantly, why does Mrs Sjoholm offer to bankroll the new restaurant on terms so advantageous to her former employees? This is not simply "providence" working on their behalf—as in *Ariel*—but Kaurismäki trying to actually play God as he had not done in the earlier films.

Among Kaurismäki's films dealt with in this chapter, *Ariel* has a happy ending but I noted the irony in an escaped felon getting away from his home country with his loved one—and with stolen money. *Take Care of Your Scarf Tatiana*, the happiest film, also had an ironic ending in which his main protagonist settles in Soviet Estonia, gladly leaving Finland behind him. It is such irony that one misses in *Drifting Clouds*. Why Kaurismäki changed track is difficult to say without becoming speculative. But I have pointed to evidence that he was a well-wisher of the Soviet system and its end could have gradually nudged him into taking a conciliatory position within Finland—when all his prior films had been implacable in their antagonism to its "capitalism by democratic consensus." He therefore distinguishes the "good capitalist" from the others—perhaps for the first time—through Mrs Sjoholm[26] and tries to see some good in the system.

Lastly, there is the dog in *Drifting Clouds* and Kaurismäki's placing of dogs in his stories has received notice.[27] Dogs appear earlier as in *Leningrad Cowboys Go America* (1989), *La Vie de Boheme* (1992) but from *Drifting Clouds* onward they seem to assume a more important position—as also in *Juha* (1999), *Man without a Past* (2002), *Lights in the Dusk* (2006), *Le Havre* (2011), and *The Other Side of Hope* (2017). They are also acknowledged in the credits:

> Kaurismäki's canine actors are usually included alongside the human actors in the films' closing credits, a conventional acknowledgement of the non-human animal's participation in a film, but in his more recent features their names also appear in the

> opening credits, a far less common practice, reflecting a more equitable recognition of their contribution, an idiosyncratic inclusivity that is subversive precisely due to its straightforward (its "straightfaced") presentation or performance of a rather radical non-anthropocentric stance.[28]

An interpretation of Kaurismäki's use of dogs is that they are incommunicative and inexpressive like his characters.[29] Since so much has been said about the dogs and they often feature in the posters, it would be helpful to study the dog in *Drifting Clouds*, a terrier that lives with Ilona and Lauri. My own response is to see the dogs in the later films occupying the position of Irmeli's son in *Ariel*. The boy (who spent his time reading comics) in that film rarely spoke but was always obligingly doing his bit, reassuringly in harmony with his mother and Taisto, for whom little went right otherwise. My reading was that the boy brought some sunshine into their difficult lives, the narrative logic being that even the most desperate lives were entitled to some of it. But the boy still had to be negotiated with—like the two having to shut him out of their room on their wedding night when he sought to come in too.

In considering *Drifting Clouds*, we find that if the dog is uncommunicative, Kaurismäki's focal characters are also more expressive than they used to be. Mrs Sjoholm, for instance, is voluble when she explains why she is funding the restaurant project. Her rescue act is implausible and Kaurismäki is perhaps defending it. The relative volubility of the characters helps to bring them closer and there is a sense of "working-class solidarity" not evidenced earlier. The dog itself is watchful as Irmeli's son was and, even if he does nothing to assist, his effect is comparable to Stanley Ipkiss's terrier in *The Mask* (1994), as an element of reassurance.[30] My complaint about *Drifting Clouds* is that there is an overload of reassurance—contributed to by the dog—since Mrs Sjoholm has already rescued the protagonists, as Kaurismäki might not have unequivocally allowed in an earlier film. These guarantees turn *Drifting Clouds* into a feel-good exercise, disappointing from a filmmaker known for more uncompromising works, made so by his irony. But it has also made him more popular with film audiences—something that a work as severe as *The Match Factory Girl* perhaps could not make him.

NOTES

1. These details are to be found in Andrew Nestingen, *The Cinema of Aki Kaurismäki: Contrarian Stories* (New York: Wallflower Press, 2013), p. 12.
2. Music for ironic effect is used by several other filmmakers covered in this book—Fassbinder, Kubrick, Altman, and Balabanov.

3. Parody is not necessarily mocking the original. "One critic makes a useful distinction between parodies that use the parodied text as a target and those that use it as a weapon [...]. The latter is closer to the truth of modern, extended, ironic parody, while the former is what has more traditionally been considered as parody." Linda Hutcheon, *A Theory of Parody: The Teachings of Twentieth Century Art Forms* (New York: Methuen, 1985), p. 52. One could propose that Kaurismäki is parodying Dostoyevsky in order to attack a milieu much poorer in thought and this goes along with his inarticulate protagonists.
4. Here is Kaurismaki on the subject: "This nation is so insecure that it's just the greatest thing ever if some foreigner says something nice about Finland, or some Finn jumps or hops or bounces farther than anybody else. In that sense, when you have some success internationally, then that success is accepted here, too. Before that, my films were part of the freak show." The author goes on to say, "The small nation sees Kaurismäki's success as Finnish success, but Kaurismäki criticises such national enthusiasm." Andrew Nestingen, *The Cinema of Aki Kaurismäki: Contrarian Stories*, p. 2.
5. As an instance, Hamlet's conduct in the boardroom is especially bizarre when he is shown sitting in a corner drawing with crayons while the board is discussing the company's business plans.
6. Andrew Nestingen, *The Cinema of Aki Kaurismäki: Contrarian Stories*, p. 32.
7. Nestingen, p. 32.
8. Nestingen, p. 28.
9. "Institutional action, suggests Kaurismäki, is another word for cronyism. This observation is the premise of *Hamlet Goes Business*, in which a powerful family corporation enjoys shelter from the law." Nestingen, p. 67.
10. "Clearly an erratic combination of serious and comic elements, *Hamlet Goes Business* transforms the main conflicts of Shakespeare's play into a present-day tale of capitalist treachery and working-class solidarity." Ulrike Hanstein, "Masquerading, Underacting and Screen Performances," in *The Films of Aki Kaurismäki: Ludic Engagements*, ed. by Thomas Austin (London: Bloomsbury Academic, 2018), p. 175.
11. Bela Tarr's Hungarian film *Satantango* (1994) deals with the same scenario differently based on the actual experience of the economic collapse.
12. Andrew Nestingen, *The Cinema of Aki Kaurismäki: Contrarian Stories*, p. 69.
13. Jaakko Seppala, "The Camera's Ironic Point of View: Notes on the Strange and Comic Elements in the Films of Aki Kaurismäki," in *The Films of Aki Kaurismaki: Ludic Engagements*, ed. by Thomas Austin, p. 92.
14. Pietari Kaapa, "Three Ecologies of Kaurismäki," in *The Films of Aki Kaurismaki: Ludic Engagements*, ed. by Thomas Austin, p. 48.
15. All countries, even if they are not democratic, claim to be democracies, which may explain that this is a convention that is hardly ever broken. Filmmakers must express faith in the system unless there is an envisaged utopia on behalf of which the crime is committed.

16. Here is an account of his transformation: "He gets swindled. He gets arrested and thrown in prison. He allows his friend Mikkonen to take a risk that leads to Mikkonen's death. Why does Taisto do all the wrong things? It is not a matter of naïve innocence and misplaced trust, for Taisto knows how to work, how to get up from a knockdown, and fire a handgun. What we have is inertia, a displaced man whose habits and ways are ill-suited to the world he inhabits, but who slowly refashions himself to inhabit a new social world." Andrew Nestingen, *The Cinema of Aki Kaurismäki: Contrarian Stories*, p. 106.
17. Nestingen, p. 27.
18. The film I particularly invoke is *Butch Cassidy and the Sundance Kid* (1969) where the outlaws escape from the United States but are killed in Bolivia, meaning that retribution cannot be escaped.
19. The author is Sergeanne Golon (Anne Golon) known for her *Angelique* novels which have sold 150 million copies worldwide.
20. Dressing up as a way of denoting class was also a strategy used by Fassbinder in *Fox and His Friends*.
21. Aarne is brutal at dinner, even as she eats her desert, "If you think there is something lasting between us, you are sorely mistaken. Nothing could touch me less than your affection. Better run off now."
22. Kaurismaki modeled his film on Robert Bresson's *Mouchette* (1967) and the girl in that film feels much of the same rage that Iris feels in *The Match Factory Girl* and is equally undirected. The difference is that Kaurismäki feels that emotion himself while Bresson only portrayed it in his character. See Andrew Nestingen, *The Cinema of Aki Kaurismäki: Contrarian Stories*, p. 44.
23. Thomas Austin (ed.), "Temporality in Kaurismäki: Anachronism, Allusion, Tableau" in *The Films of Aki Kaurismäki: Ludic Engagements*, p. 23.
24. Andrew Nestingen, *The Cinema of Aki Kaurismäki: Contrarian Stories*, p. 88.
25. Here is a contrary view: "The discourse of Finnish drinking has been contested by different quarters in line with diverse political agendas. Nineteenth- and twentieth-century conservative elites construed drunkenness as symbolic evidence of the rural masses' lack of development and sophistication, establishing drunkenness as a justification for coercive forms of social reform [...]. Kaurismäki's generally happy drinkers contest the premise of these discourses, rejecting the construction of alcohol consumption as lacking sophistication, self-control, and middle-class manners. Kaurismäki rejects the model of the problem drinker, replacing it instead with the stylish drinker. His characters drink in different ways, but alcohol is almost always represented as involving conventions of consumption and sociability that maintain control over alcohol intake [...]. For example, the alcoholic cook Lajunen in *Drifting Clouds* loses control, but his colleagues look out for him and protect him from himself. These representations reject the problem drinker model." Nestingen, p. 76. A question that could be put here is whether this does not conflict with alcohol being presented as a problem within the family when Lauri turns to alcohol when he is unemployed and causing grief to Ilona.

26. Kaurismäki creates the setting right at the beginning when she is made someone who cares for her restaurant and her employees. Lajunen is not sacked after his outburst but given a final warning. But her reappearance as savior is nonetheless abrupt. “During the nostalgic tango on the last evening at Dubrovnik, there is a track in to a close-up of Mrs Sjöholm, giving gentle emphasis to her sorrow. She then disappears from the film only to reappear like a deus ex machina when all hope for the protagonists, Ilona and Lauri, appears to be lost.” Henry Bacon, “Levels of Typification in in Aki Kaurismäki’s *Drifting Clouds*,” in *The Films of Aki Kaurismäki: Ludic Engagements*, ed. by Thomas Austin, p. 158.
27. Michael Lawrence, “Deadpan Dogs: Aki Kaurismäki’s Canine Comedies,” in *The Films of Aki Kaurismäki: Ludic Engagements*, ed. by Thomas Austin, pp. 187–203.
28. Lawrence, p. 193.
29. Lawrence, p. 187.
30. At the very least, the dog reassures that Ilona and Lauri will stay together since they are both attached to him. As comparison, a dog would have been completely out of place in *The Match Factory Girl.*

7

Aleksei Balabanov: Collapsed Utopia

Aleksei Balabanov, like many of the best Soviet filmmakers of the Brezhnev era and after, is not as well-known internationally as he deserves to be. In fact, Soviet/Russian achievements in the field of cinema from the Stalin era onward are represented in film history only by a handful of filmmakers, with Andrei Tarkovsky as the most widely celebrated. In terms of craft, the USSR, which had been at the forefront of filmmaking in the 1920s, continued to excel because of the film school (VGIK) in which the best filmmakers taught; but Soviet films do not feature significantly in the global cinephile's consciousness as they might have. A reason is the isolationism of the Soviet Union when only a small number of films like Mikhail Kalatozov's *The Cranes Are Flying* (1959)[1] received notice at western film festivals. Tarkovsky and Sergei Paradjanov (*The Colour of Pomegranates*, 1969) were exceptions since the proscription of their films within the USSR[2] also contributed to their fame in the West. In the chapter on Fassbinder, I felt that some knowledge of German cinema before the war and after was necessary to historicize the filmmaker's traits and that is also true to a limited extent of Balabanov's films since, as in the case of Germany, the USSR also experienced political change that influenced the films. Before actually examining his films, I will look briefly at some relevant aspects of Soviet cinema from the 1920s onward, focusing on aspects pertinent to satire and Balabanov.

Montage, choreography and polyphony

The key notion at the heart of early Soviet cinema is montage although each filmmaker-theorist—Sergei Eisenstein, Vsevolod Pudovkin, Dziga Vertov, and Kuleshov, to name the important ones—approached it differently. The key notion underlying the montage theory of the early Soviet era is that meaning only arises when individual shots are juxtaposed. Since the effects thus obtained were striking, one would have expected montage theory to grow from strength to strength. But the filmmakers abandoned it without a theoretical basis being offered in the 1930s.

At the beginning of Pudovkin's *Mother* (1926) is a sequence in which the drunken father returns home and tries to appropriate the clock—to pawn it for more vodka—while his wife and son sleep. While the father's movement in this sequence heightened by the relationship with the two others is caught through montage, the overall effects are obtained by combining concordant elements in the manner of orchestrated music. This is even truer in the sequence that follows—a pub scene in which the workers are "resting" after work. It is important here that Pudovkin also has the means to convey group movement without using montage because this means that such movement and not the cut is the primary interest. To elaborate on this, in the pub scene there are inclusive shots in which activities of different kinds are happening at the same time—a waiter serving someone, a drunkard staggering from his chair toward the counter, an orchestra playing in the top left-hand corner, people dancing in the far distance to the right.

These inclusive shots are immediately cut to tighter others in which details are singled out but the presence of the inclusive long view is unmistakable—as is Pudovkin's orchestration of movement within a single frame. When Pudovkin shows two or more faces within a single frame, the faces either represent different character types or are composed differently—like the three men at a table watching a disturbance, each one studying it with a marginally different attitude—attention, irony, and indifference. That this is "composed" like music—with harmony as the operating principle—and is not simply a "realistic" depiction of an actual scene is suggested by the fact that DW Griffith—who influenced the Soviets greatly—does not have the same capabilities at his command. *Birth of Nation* (1915), for instance, handles its action very differently. Griffith's film rarely has two or more faces within a single frame and groups are shot from a distance without "choreographing" crowd movement. In an early segment dealing with a flock of abolitionists, the seated group is filmed from behind so that we do not catch individual faces; the only countenance—seen from a distance—is that of the man getting contributions.

Griffith is making us absorb the film at the plot or story level and empathize while Pudovkin is making us aware of its construction. Pudovkin's deviation from Griffith arises because, like the Constructivists and Bertolt Brecht in later theatre, the Soviet filmmakers of the 1920s tried to create an audience that would be one of active viewers. If there is an attempt here to mobilize the public toward a single political agenda, "plane polarize" it as it were, there is also an admission that the public is initially in an unpolarized and unreceptive state from which it needs to be critically aroused, and this is where "choreography" becomes important; its application or employment conveys the sense that the world has no single meaning and that perceptions of its nature are "polyphonic"—though ideological affinities could help subordinate the different perceptions. To draw a literary parallel,

Mother is like a "dialogic text" in which the constitutive elements are allowed liberty/polyphony which, although subordinated to the authoritative discourse of Marxist-Leninist doctrine,[3] still reveals itself above it.

Mark Donskoy's remake of *Mother* (1956) has an early segment comparable to that from Pudovkin's film—the belligerent father returning to his home and clashing with his wife and son. In Pudovkin's film, the father has his eyes only on the clock and the iron that he wishes to take away and he notices his wife and son only when they try to stop him, and he pushes them aside. In Donskoy's film, there is hostility between the man and his family but his acknowledgment of them comes first; it is in their respective use of the eyeline match that the attitudes of the two films are largely manifested.[4] When people make eye contact conspicuously, there is an acknowledgment of each other not only as persons to negotiate with but also as viewpoints to accommodate.

Where in Pudovkin's film the mother and her husband exist on different planes of understanding but are forced to transact because of the mother's instinct for survival, Donskoy's film shows man and wife, despite their underlying distrust of each other, as being together on the same plane—i.e. sharing a single understanding of their milieu. The difference between the two couples is this: the first is just two people who share little (Pudovkin) while the other is mutually distrustful but with the cause of the distrust *established* between them (Donskoy) and only the first may constitute "polyphony" in the sense of multiple independent voices. If the quality of "polyphony" is in evidence in Pudovkin's realism but not in Donskoy's,[5] the Stalin era is the period in which the polyphony of Soviet cinema was interrupted—until it re-emerges in later cinema in the Gorbachev era. Stalinist ideology could not countenance differences within the working class.

Satire in Soviet cinema

The notion of the polyphonic text that I just invoked owes to Mikhail Bakhtin, a Soviet literary critic and philosopher who studied the novel. Bakhtin did not restrict his study to Russian authors but also examined the works of writers like Charles Dickens (*Little Dorrit*) and concluded that a literary work is a site for the dialogic interaction of multiple voices which are the product of "multiple determinants specific to classes, social groups and speech communities."[6] "Polyphony" suggests the interplay of different kinds of speech within a single verbal text but it is difficult to similarly conceive of different kinds of "speech" in visual narration. But a film is narration as much as a novel is and there should still be a parallel of some sort. Since the teleology of a fiction film corresponds to "intent" in authorial narration—because it is the course imposed on the narrative by the director of the film—the resistance to the narrative's teleology, i.e. the interruptions in the narration and the elements that are not "resolved" will constitute the polyphony. As an instance, the trajectories of major characters *not* tied to the resolution of the story could represent the polyphony element. An immaculate plot, it may be noted, would eliminate these elements and "plane polarize" the text.

Satire, I propose, depends on "multiple voices," and an illustration would be the overlapping dialogue in an Altman film, which serves that purpose. Luis Buñuel also got his effects by showing people thinking at cross-purposes and not fully comprehending what drove the other, as for instance in *The Diary of a Chambermaid* and *El*. The perceived "stupidity" of Fassbinder's characters (as I described it) owed to their insensitivity to the "larger picture," implying self-involvement and lack of mutual understanding. Emmi did not understand her own political errors in *Fear Eats the Soul* and repeated them with Ali, not understanding his position. Kubrick's *The Shining* is focused on the thinking gap between Jack Torrance and his wife Wendy.

The satire/irony in Russian literature of the nineteenth century can be found in the writings of Nikolai Gogol and Fyodor Dostoyevsky. The multiple voices in Dostoyevsky's novels often produced satirical elements as in *The Possessed* and short fiction like *The Ugly Story*, filmed in 1966 by Alexandr Alov and Vladimir Naumov. Since these works of fiction—like those of Nikolai Gogol—were set in Tsarist Russia, some were adapted as films in the USSR. But satire set in the contemporary Soviet Union was rare even in the thaw period (under Nikita Khruschev), not surfacing until the glasnost era. A celebrated satire of the thaw period is Elem Klimov's *Welcome or No Trespassing* (1964) about the restrictions in a children's camp but it becomes a comedy about the children *coming together* and getting the better of the figure of authority. Among the first robust satirical films was Vladimir Bortko's *Heart of a Dog* (1988) based on Mikhail Bulgakov's novella, a mock "sci-fi" story set immediately after the revolution. The 1980s can be understood as the period when polyphony returned to Soviet cinema. The overall sense to be got from Soviet cinema earlier to that was of dialogues concluded because "everything had been resolved," as hinted in a great Soviet film of the 1980s.[7]

The films of Aleksei Balabanov

Aleksei Balabanov has the paradoxical distinction of being considered both a nationalist and the maker of "negative things."[8] The nationalist side is evidenced in his creating a new Russian hero for the post-Soviet era in films like *Brother* (1997) and *War* (2002)[9] but this chapter pertains to the other films about "negative things" which gives him greater importance. Dostoyevsky and Gogol too, while excelling at satire, had strong religious or nationalist sides to their work.

Born in Sverdlovsk (Ekaterinburg) in 1959, Aleksei Balabanov (who passed away in 2013) graduated as a military translator in 1981 and served in Africa and the Middle East. Returning to his hometown he worked from 1983 to 1987 as a director's assistant at Sverdlovsk Film Studio, then entered the Graduate Courses for Scriptwriters and Directors (VKSR) in Moscow, where he studied in the Auteur Cinema experimental workshop of Lev Nikolaev and Boris Galanter (Director's Sector). Graduating in 1990 he returned to Leningrad to shoot his first feature film, *Happy Days* (1991), supposedly an adaptation of Samuel Beckett's play,[10] and the product of Lenfilm's Studio of First and Experimental Films with which he was associated during Aleksei German's tenure (1988–92). *Happy Days* was followed in 1994 by *The Castle* (based on Kafka's novel), often considered Balabanov's weakest, including by the director himself. In 1994, together with fellow director-producers Sergei Selianov and Viktor Sergeev, Balabanov founded the St Petersburg production company CTV, with which his work has been closely associated.

Balabanov's next work, "Trofim," was one of four short works that together compose the compilation film *Arrival of a Train* (*Pribytie poezda*; CTV, 1995), shot to celebrate cinema's centenary. Balabanov's contribution was hailed as the best of the four shorts, but it was his next work, the action film *Brother* (1997) that marked the start of his visibility. *Brother* was a hit both at the box office and on video. It was followed by one of his most controversial films, the stylized historical drama *Of Freaks and Men* (1998), which, despite winning the Nika award for Best Director and Best Picture in 1998, encountered severe distribution problems both in Russia and internationally because of its disturbing content, such as the shots of two naked children, the film's conjoined twins. As a critic put it, "I am very glad that I live in an emancipated country where the highest cinema prize is given to a film for which broad distribution (and correspondingly popular success) is impossible by its very definition."[11] But the controversies did not die down since *Cargo 200* (2007), which actors declined to be in and that some theatres refused to show, remains one of the most disturbing films ever made. Balabanov has worked in a variety of genres including the arthouse film, action, and romance and I have selected only six of his films for this chapter, broadly those that fit the aims of this book.

Happy Days (1991)

The film is described as an adaptation of Beckett's play but it bears no discernible resemblance. Beckett's play was focused on a woman buried up to her waist in mud, and later up to her neck, being optimistic and cheerful. In Balabanov's film, the protagonist has no name and has forgotten the one he had once been given on account of a head injury. He has no place to stay and the city (St Petersburg) is a ramshackle space, no people about on the streets, and only an occasional empty tram providing a sense of movement and life. He searches for accommodation around the Winter Palace, finds lodgings where the landlady calls him Sergei Sergeevich but her husband, the real Sergei Sergeevich (a collector of antique clocks) appears and evicts him. Another potential landlady gives him the name Borya. In a cemetery where he spends his time, he meets a girl who lives in a dilapidated mansion, claims German aristocratic ancestry, and works as a prostitute. Eventually, the protagonist finds shelter in an abandoned rowboat near the river, which would float if the city became waterlogged. The film uses an American jazz track "Too many tears," a bit of Wagner heard through windows of houses and a tune out of a music box carried by the protagonist all of which suggest nostalgia.

The film is exquisitely filmed in black-and-white and the city's decayed ambience should evidently be a key factor in the film's meaning. A factor is the protagonist conceived more as a cartoon character than an actual person since he wears his hat even when he is sleeping.[12] In tune with this portrayal, the others who appear in the film are also fleeting caricatures. This means that the protagonist need not develop or transform in the course of the narrative into something else. It could also be argued that the emptiness of the city is resorted to since it would be difficult to film in a populated space and still reduce everyone to the level of a cartoon. Two animals, a donkey and a hedgehog, also feature and only they are allowed to exist in themselves. The protagonist is arguably reflecting the citizen in 1991 with Russia at the crossroads. There is no reference to the country's recent past and the city's architecture only recalls its imperial period. The protagonist, because he has no memory even of his name, is someone cut off from this past, with nothing to replace it in his consciousness. A home represents a past especially in a locale as full of its traces as St Petersburg and he is perpetually homeless.

I am not certain that the film succeeds, but it can be interpreted as a nationalist lament at Russia's cultural state; there is a sense in the film of the Communist past not having added anything of value. It is as though an empty period separated St Petersburg's imperial past from the present when everything simply decayed.[13] Communism in the USSR was once expected to transform people to bring them to a new state of social consciousness but where they stand or will develop into remains unresolved, which is why the characters are out of a cartoon that cannot develop into anything definite. Balabanov takes only Beckett's title from the play—ironically since the fall of Communism was being celebrated worldwide when the film was made. A hat is also something associated with the attire of the woman in Beckett's play, especially because of its incongruity with the circumstances it depicts. In Beckett's play, it also indicates an unchanging good cheer of the wearer regardless of the immediate circumstances.

Happy Days can be termed ironic but its satirical purport is not self-evident and that remains true of Balabanov's best films—that not only invite interpretation but also refuse to submit to it. The donkey and the hedgehog for instance cannot be interpreted exactly but they are both symbols of peace. The donkey is the opposite of the war horse and there is even a fable by Aesop comparing the two. St Petersburg played a key historical role in the revolution and the presence of the two passive creatures in Balabanov's film suggests that history has played itself out. It conveys the sense of the meek having inherited the milieu after the strife had concluded.

Of Freaks and Men (1998)

This film is also set in St Petersburg but at the turn of the century and filmed in sepia tones to mimic silent cinema with inter-titles. Between *Happy Days* and this film, Balabanov has gained enormously in confidence; the film is brilliantly filmed and acted, highly original, and extremely disturbing. The film deals with two respectable families, that of the widowed engineer Radlov who lives with his daughter Liza (Dinara Drukarova) and maid Grunia and Dr Stasov, who lives with his blind wife Ekaterina, maid Daria, and adopted Siamese twins Kolia and Tolia. The two families are linked by a group of criminals led by Johan (Sergey Makovetskiy). Johan is Grunia's brother and a pornographer with a photo studio; he has an assistant Victor Ivanovich (Viktor Sukhorukov) and uses the services of a photographer named Putilov. The pornography produced by Johan is limited in its scope and shows young women being spanked on their bare bottoms as punishment for unspecified misdemeanors. Both respectable families get into the clutches of Johan's gang. Engineer Radlov predicts that cinematography will be the new art, the greatest ever, and the film shows the pornographers moving from still photography to cinematography but filming the same flagellation of bare behinds by a nanny.

At the basic level, the film is satirizing the early Soviet expectations of cinema, mocking its ideological puritanism. Lenin, it may be recalled, said, "Of all the arts the most important for us is the cinema." Balabanov is proposing that, as for most other widely applicable technologies, the first use of cinema was pornography. In itself, this may not be novel as a statement but given the reverence with which cinema is treated, this is deeply ironic. The pornographers use the same basic gag—punishment by spanking—even with the arrival of cinema.

The two gentle families are destroyed by Johan when Engineer Radlov and Dr Stasov are killed and the innocents fall into the grasp of the pornographers. But they are weak and riddled with perversities and strange passions. Stasov is hated by his blind wife, who yields to the voice of Victor Ivanovich, with whom she "falls in love." Victor Ivanovich using the tip of his umbrella to inspect the woman's private areas is disturbing but it excites her. Radlov is smitten with his maid, while his virginal daughter secretes flagellation photos away among her clothes.

The Siamese twins Kolia and Tolia are also exploited as performing exhibits, they are made prey to drink and their own naïveté. The twins sing together and they are also made to perform before a high society gathering. Johan gets his own former nanny, a decrepit old woman, who is the dominatrix in the flagellation episodes filmed by Putilov, who loves Liza secretly. The nanny dies when the exertions of acting prove too much. Victor Ivanovich is shot dead by the twins and they head east, from where Dr Stasov came. Kolia, who also loves Lisa, has a sexual relationship with her, her second after Victor Ivanovich. Lisa eventually heads west, which has been her ambition all along. Their living room overlooks a railway station and an engine is seen being watered by Lisa. The reference (which works as leitmotif) may be to the Lumiere brothers' famed documentary about the train.[14] Later when Lisa fulfills that ambition by traveling west, she spies a transvestite in a window, recalling those in the Reeperbahn, and gets herself flogged. The twins travel east and succeed in performing exhibits but Tolia dies abruptly, decreeing Kolia's fate. The film ends with Putilov, now a famous film director in the west; Johan is in the cinema watching Putilov's film *Punishment for a Crime: A Real Life Drama*, moved to tears at the last images of his dead nanny administering beatings to Liza's behind.

Of Freaks and Men is exquisitely filmed and its visual lyricism belies its actual content. Balabanov paints an extremely perverse world matter-of-factly and it is only the film's visual beauty that distracts us. The film's narrative, for all the perverse exploitation it portrays, is not neatly divided along "good" and "bad" lines. Johan and Victor Ivanovich are frightening creations and the cold grin that the latter flashes on gaining entry into the two respectable households is unsettling. Johan, for his part is impassive, whether his intention is to propose marriage to Lisa or have her flogged. At the same time, the victims become dependent on the abusive relationships they are embroiled in, and the men who should protect them are ineffectual. Giving the victims strange private passions also makes the film polyphonic. The film can be read as an acknowledgment of the irrational human impulses that made the Communist utopia subsequently fail. In other films, we will see alcohol and drugs playing the same role.

The film works at several levels and, as already indicated there is, first, the reference to cinema. Apart from the cherished works of the pioneers that the film is parodying in this fashion, there is also the well-known puritanism of Soviet cinema. The Soviet authorities also found it inconvenient to allow the portrayal of interpersonal conflict since they professed to have already created an ideal society where such conflict had been fully eliminated and no one could be truly culpable of serious social errors. War films like Elem Klimov's *Come and See* (1985) rely on the external enemy (Germany) to portray violent doings and express the anger otherwise rarely in evidence. Balabanov's film can be read as a subversion of the earlier Soviet cinema; innocence is as scarce in it as mischievous intentions were in the earlier films.

The second aspect is that the film came out in the Yeltsin era in which organized crime gained control over the economy.[15] One might anticipate how this might have manifested itself in the everyday space with the cultural elite and the *nomenklatura* (the political elite) progressively helpless against criminal elements, represented in the film by Johan and Victor Ivanovich. Russia had not only a very cultured elite class but also an uncontrollable proletariat with ambitions to move upward after the fall of Communism. The Communists idealized this unruly class as revolutionary but it has been argued that it was anarchy rather than foreign imperialists that the Red Terror under the Bolshevists had to contain.[16] That the cultural elite is helpless against criminality is ironically suggested by Balabanov's use of Prokofiev and Mussorgsky on the soundtrack,[17] and showing St Petersburg's splendid buildings hiding murky secrets.

The criminal class is portrayed in a major part of Balabanov's subsequent films and what has just been said about *Of Freaks and Men* will hopefully shed some light on its significance. The next film to be examined *Dead Man's Bluff* (2005) begins in a classroom in the new millennium where the teacher describes the building of startup capital in the 1990s through "crime merging with authority."

Dead Man's Bluff (2005)

The story of the film is an illustration of how people accumulated "startup capital" in the 1990s and is a satirical gangster comedy with two hit-men Sergei (Alexei Panin) and Semyon (Dmitri Dyuzhev) who work for a crime boss Sergei Mikhailovich (Nikita Mikhalkov). The film begins in a room filled with naked bodies and a man with a syringe preparing to kill a gagged person, a gangster. Three masked men break in and there is only one survivor in the shootout, Stepan (Viktor Sokhorukov) who kills his remaining associate, another of the three masked men.

The film contains elements reminiscent of Tarantino's *Pulp Fiction* like having two men although they are not articulate like Jules and Vincent but loutish. Stepan, it turns out, is a crooked police officer and his group is trying to snatch money from the protagonists. Sergei Mikhailovich is a brutal master but a loving father who dotes on his son. The chamber with the naked corpses and Sergei Mikhailovich conducts his young son there asking him not to get his shoes dirty since there is blood on the floor. Evidently, he is preparing his son for a life of responsibilities.

Sergei and Semyon have some heroin belonging to their boss stolen by Stepan and his associates, but they manage to retrieve it after killing their adversaries.

The title of the film comes from the game they make the other group play, something generally known as "Russian Roulette." Sergei has a bullet in his gut and a medical student is called in to remove it. This part of the film is reminiscent of *Reservoir Dogs*. Instead of making the heroin over to Sergei Mikhailovich, they use it as startup capital in Moscow and end up as parliamentarians with an office in the vicinity of the Kremlin, and Sergei Mikhailovich is now their security guard.

The film may not sound overly funny as just narrated but the performances and laconic filmmaking style ensure it, unless one is appalled by the casual violence and the gallons of blood. Another aspect is the fun made of a darker-skinned character called "Ethiopian" by his comrades. This is not Balabanov being racist but showing the racism prevalent in the criminal class without taking the expected ethical position on it; it would perhaps be ludicrous to select such characters to portray and have qualms about their portrayal. Regardless of the horrific violence, therefore, Balabanov's tone remains ironic and hence tries to shock less than Tarantino.[18] The fearlessness of the characters in the face of death, for instance, is on account of their obdurate stupidity and those who survive duly become parliamentarians in Moscow. This can be interpreted as a satirical take on the game of "Dead Man's Bluff" used in the title.

Still, it is more like a local joke about economic issues in Russia with leading actors playing bit roles. Nikita Mikhalkov is a well-known Russian filmmaker (and actor) with an Oscar for Best Foreign Language Film (*Burnt by the Sun*, 1995). Lead actors doing bit roles of the lowest characters unselfconsciously adds to the humor. But since Balabanov often works in the popular idiom, the film is not demanding that its satire be taken as mordant. The best thing about the film is perhaps that it is structured like an economic explanation of Russia's present. It has been noted that the opening lecture scene of *Dead Man's Bluff* mocks Russia's incapacity to account for its own economic conditions.[19] The notion of "crime merging with authority" is, for instance, illustrated simply by one of the gangsters Stepan actually being a policeman. It can be seen as a commentary on the Putin era but that commentary is not sharp.

But *Dead Man's Bluff* gives us no inkling of the next film to be examined, *Cargo 200* (2007) which is a profoundly disturbing look at the USSR in the last stages of its collapse in 1984, with the war in Afghanistan still raging. Some people apparently walked out during the film and pregnant women were warned to leave the cinema for the shock that the film could cause.[20]

Cargo 200 (2007)

The film claims to be based on a true story set in 1984 but it is more likely an adaptation of William Faulkner's novel *Sanctuary* (1931). It begins with a conversation between two brothers bemoaning the ways of the younger generation. Artem (Leonid Gromov) is a professor of scientific atheism at the University of Leningrad on his way to Leninsk to see his mother while Mikhail (Yuriy Stepanov) is a colonel in the army who, in a low voice, is explaining how things have gone after Yuri Andropov's death.[21] Mikhail has a daughter Lisa who appears with her boyfriend Valera (Leonid Bichevin). Mikhail quietly wonders how Valera can continue as he does; his antecedents are unknown, he lives with an old woman claiming she is his granny but earns thrice as much as Mikhail. Artem himself has an unruly son named Slava who he got out of military duty but instead of studying, the boy parties most of the time.

Lisa and Valera, along with some others, are leaving the next morning for a friend's dacha. It is evening when Artem and Valera leave separately; Artem's car breaks down on the way outside a village named Kalyaevo. Valera goes to a party at Kalyaevo, drinks heavily, and meets Angelika (Agniya Kuznetsova), whose dacha it is that the youngsters are driving to the next morning. Angelika is the daughter of the Secretary of the District Party Committee. Artem meanwhile needs to get his car going and finds a farmhouse, that of a moonshiner named Alexey (Alexey Serebriakov).

This is where a disquieting element is first introduced since the man who directs him inside is frightening in his intense impassivity. There is something reminiscent of *The Texas Chainsaw Massacre* (1974) at this juncture in the way the bootlegger Alexey is introduced, although the film does not develop into full-fledged horror and Alexey only starts an argument with Artem. Artem needs his car repaired and Alexey sets his Vietnamese handyman Sunka (Mikhail Skryabin) the task but offers spirit to Artem meanwhile, who gets drunk although he keeps up the argument. The argument between Alexey and Artem is about God and Alexey believes in a Kingdom of the Sun, while Artem is rational with agnostic explanations for the "conscience."

The men are drunk but what gradually emerges is the impossibility of a rational society along Soviet lines when the people meant to constitute that society are irrational; would a constructed society not need to take that into account? Such an endeavor might be like making a picture from nature—where there is scarcely a straight line—through an assembly of matchsticks. But Artem's car is soon repaired and he departs though Sunka warns him that it might not make it to Leninsk.

To Alexey's farmhouse also come Valera and Angelika; though Valera is already drunk, he still wants moonshine. Valera leaves Angelika in the car but the expressionless man encountered by Artem now comes up to the car and peers at Angelika who goes into fits of screaming. Valera has turned on the music in his car, some noisy Russian pop that sounds like growling, and the atmosphere is horrific although the strange man leaves Angelika and goes back inside.

The film has different aspects that need to be noted and the first is its association with horror. The horror film is a genre with conventions that are not always easy to identify but it can perhaps be asserted that horror films (as opposed to those that inspire only terror)[22] deal with the notion of contamination in some form or another. *The Exorcist* (1979) and *The Omen* (1976) are about the contamination of the Christian world by dark/pagan forces, while *The Fly* (1986) is about the contamination of a human body by genetic material from a foreign organism. But there is still another sub-category exemplified by *Psycho* (1960), *Deliverance* (1972), and *The Texas Chainsaw Massacre* (1974), which are all horror films. These films are not apparently about contamination, at least in the same sense, but they still involve the notion though in a modified way. The films tell the stories of one person or a group of (usually helpless) people who venture into terrain away from the civilized mainstream and are subjected to bizarre and horrific acts by people nominally civilized but who have allowed civilization to slip away; the protagonists are initiated by these "mutants" into horrific reality.

What these films also have in common is that no explanation is provided for why these "rejects" of civilization have mutated. One rational/social science way of interpreting the kind of horror film to which *Psycho* and *The Texas Chainsaw Massacre* belong (films about a terrible initiation) in the light of the contamination criterion offered is that civilization and organized society are never able to achieve their ends because of the dark human psyche and that they unwittingly create cesspools of disaffection in which breed social mutants of various sorts. To provide an analogy, one is familiar with scum collecting in the nooks bordering fast-flowing rivers and these nooks provide shelter to creatures that cannot survive in the flowing current.

In *Cargo 200*, Angelika takes the place of these "innocents," but she is, ironically, the privileged representative of a system in collapse. She undergoes the experiences not in a remote location but in familiar terrain. The slumbering state and the Party have become so alienated from the society of their own creation that social mutants thrive under their noses. Balabanov staunchly refuses to take sides—between the innocent victims and the perpetrators—because the victims in this case are implicated in the creation of the monsters. Angelika is subjected to abuse after abuse (to which she submits) and one scarcely recalls a woman character in cinema subjected to so much, and this certainly creates difficulties for the spectator. Still, she is not only "woman"—which might make the film unforgivably misogynistic—but also the softened product of a privilege-breeding system, the provincial *nomenklatura*.

The film takes the shape of a thriller in the sense that much is revealed only in stages. The man with the impassive countenance at the moonshiner's place and who kidnaps Angelika (after killing Sunka) is Captain Zhurov (Alexei Poluyan) from the militia (police) and he has "fallen in love" with her. He takes her home as his "wife," and chains her to a bed while his drunken mother watches television, initially some state proceedings from Moscow and then Cuban dance with performances by "blackies." Here again, is the motif of irrational racial prejudices thriving in a system making rational political alliances. Angelika threatens Zhurov with dire consequences from her father and her fiancé Sergeant Gorbunov, a paratrooper in Afghanistan. But as fate has it, Kolya Gorbunov returns in a lead-lined casket (*Cargo 200*) and Zhurov takes charge of his corpse, lifting it onto Angelika's bedside, uniform, medals, and all, and waits for her to wake up from her vodka-induced slumber, to be confronted by his limp visage.

Zhurov, even while holding Angelika captive, is dealing respectfully with her father and Balabanov is commenting here about the Party losing control over the state machinery, which is housing mutant creatures like Zhurov. Alongside, as a powerful backdrop, are the trains and trolleys interminably on the move, although the streets are empty, a metaphor for mechanisms set on an aimless course indifferent to those they are ostensibly for. The film's "Leninsk" is also gargantuan but a seemingly unpopulated industrial hub. When Zhurov in his office gets a call from a woman alleging domestic violence, he and his men proceed there and find three ragged, drunken workers playing cards. The man of the house—a worker after a shift—abuses Zhurov, who responds by shooting him and throwing him in the lockup for more brutality.

Also important in the film (in a subterranean way) is the relationship between Alexey and Zhurov. Alexey has been in jail for murder and he owes Zhurov a favor that even his wife Antonina (Natalya Akimova) is ignorant of. This suggests the convoluted justice system that allows personal deals outside legality almost religious in strictness. In his argument with Artem, Alexey admitted his guilt for having killed someone in a fight for which he turned himself in, and he repays Zhurov by taking the blame for Sunka's killing. He accepts the death penalty without protest on the condition that Antonina be left alone. It is therefore left to her to establish the moral equilibrium by killing Zhurov. Angelika's abuse has been troubling but Balabanov tries to make amends this way.

The filmmakers covered in this book have often been disturbing but Balabanov stands out even among them. *Cargo 200* is a merciless film although it contains little explicit violence; making it especially disconcerting is that it zeroes in on the metaphysics of a political collapse—rather than on the failings of a political system like many East European films such as those of Andrej Wajda (*Man of Marble*, 1977) or the Romanian Lucian Pintilie (*Reconstruction*, 1968). The film can be likened to Kubrick's work in the sense of a human system deemed perfect failing but it goes beyond it by not seeing human beings only as destructive but also full of longings harmful even to themselves. Each frame emphasizes degeneration and, if this is evidently the portrayal of a system in collapse, it is as though the collapse was inevitable. Lending the film an ironic edge is Artem who offers an intellectual framework for materialism, reasonable in itself but hopeless as the instrument through which the USSR tried to engineer a rational socio-economic system into existence. Alcohol is Balabanov's metaphor for that aspect of human beings impervious to rationality. The film moves to an appropriate note of resignation when Artem visits a church and asks what "procedure" should be followed if he wished to be christened. *Cargo 200* however concludes when Valera meets Artem's son Slava at a disco and the two discuss starting a business together suggesting the Yeltsin era, touching upon something already commented upon in *Dead Man's Bluff*.

Lastly, Balabanov uses Russian pop extensively in the film when people are driving—between interludes of usually ghastly action. Particularly striking is Zhurov on his motorcycle, with the chained Angelika in his sidecar, riding impassively in the empty industrial landscape, fervent pop on the soundtrack. *In Dead Man's Bluff*, he uses bars of instrumental music the same way. It could be argued that pop not only dramatizes and raises our expectations but also makes us conscious of its inappropriateness in the context of what is happening. This is not quite like Kubrick using Rossini during a rape (*A Clockwork Orange*), which is also "inappropriate." The fact that it is *not an imposition* upon the dramatic action but fills only the interludes should be noted. It is used to influence our anticipation of the coming action, infuse it with irony instead of emphasizing the disjunction between the action and its treatment. Our reflection on what has passed is tinted by a sense of the fierce volatility in interpersonal relationships when the social structure is meant to *spell rationality and order*.

Morphine (2008)

Morphine is based on autobiographical writings by Mikhail Bulgakov and is about a doctor in a distant province in 1917 who becomes addicted to morphine. The year chosen is the year of the Bolshevik Revolution. This is a different kind of film from *Cargo 200* and is closer to *Of Freaks and Men* in using inter-titles and mimicking silent cinema. While dealing with surgery graphically, it is more episodic and is visually beautiful, also like *Of Freaks and Men*, though in de-saturated color.

In the film, Doctor Polyakov (Leonid Bichevin, unrecognizable from *Cargo 200*) arrives at a village to take up the position of the physician in charge after the previous one left at the time of the February revolution. He is treated with great respect, and among his colleagues are the nurse Anna Nikolayevna (Ingeborga Dapkunaite) and the paramedic Anatoliy Lukich. His very first patient is a man in the advanced stages of diphtheria, frothing at the mouth, and Polyakov, administering mouth-to-mouth respiration, cannot save the man. Since he fears infection, he is administered a vaccine but it sets up an allergy; he gets a shot of morphine to quell the symptoms.

It is virtually a given that Polyakov will commence a sexual relationship with the nurse Anna Nikolayevna since there is no buildup of the relationship through "negotiations" such as courtship—as would be the case in western cinema—and the sense obtained is of a hierarchical and patriarchal society in which people of standing (like a doctor) naturally enjoy such privileges. Social hierarchy is evidently a key issue since the sexual relationships outside marriage in *Of Freaks and Men* between upper-class women and men of the lower order are "violations." At the same time, Polyakov evidently owes the nurse some loyalty since he hides another liaison from her. The relationship heightens the sense of the willing submission of some classes to others. The doctor is also called "your majesty" by the others in his rural hospital.

After another medical crisis, Polyakov gets the nurse to administer another dose of morphine and that makes him addicted. Balabanov is graphic in his depictions of surgical procedures—childbirth, the amputation of a mangled leg and a tracheotomy are the key ones—and the overwhelming physicality this imparts has us wondering about Balabanov's intent. When Polyakov's first patient dies of diphtheria, the doctor and his paramedic sit by the dead man and the corpse's discolored buttocks are exhibited to us, the men smoking on either side. Rarely has a surgical procedure (if ever) been shown like this in cinema and it is compounded by the cries of the relatives, who often throw themselves at the doctor's feet. While its intent is not transparent, I propose that the raw physicality should be seen alongside the film's political thread.

As already indicated, the Bolshevik Revolution constitutes a subtext in the film. Two miles from Polyakov's hospital are some affluent landowners and Polyakov takes shelter with them in the middle of a blizzard. His companion from another hospital (an intellectual) who the landowners take to be German is a christened Jew. The Revolution being associated with Germany is a reference to VI Lenin's well-known connections. The Jew is a member of the RSDWP (Russian Social Democratic Workers' Party), the parent party of both Bolsheviks and Mensheviks, and the landowner is amused that he does not look like a worker. Later, after the Bolshevik Revolution, the landowner's residence is burned down and severely burned people are brought to be treated. But Polyakov has used up all the morphine and the substitute he has replaced it with kills them. Polyakov fears punishment but Anna Nikolayevna assures him that they are kulaks and there is no punishment for killing them. Lastly, when Polyakov finally goes to a nearby town to get a cure for his addiction, there are Bolshevik guards about, illiterate louts drunk with power but with no political understanding of the situation.

I earlier noted the reappearance of polyphony, the sense of people not interacting on a single plane, in the Gorbachev era. In early Soviet cinema, there was the created sense of workers brought to a politically receptive state where they might be able to participate in revolutionary action and partake of the same destiny. In *Morphine*, the participants in the action cannot be brought to the sense of a "common destiny." This is where the film's overwhelming physicality fits in. Most of the patients are peasants and poor people and the procedures they are made to undergo inflict enormous suffering on them but Balabanov places emphasis on their bodies rather than on them as "people,"[23] which detracts from the sense of the revolution being "willed" by people like them. The morphine in the film also plays the same role as the alcohol in *Cargo 200* (and the sexual quirks in *Of Freaks and Men* that the criminals take advantage of) —as an impediment to rationality and a symptom of the un-governability of human impulses to a rationally devised system. The silent film format also helps at this point by recalling early Soviet cinema. To cinema in general, the films of Eisenstein and Pudovkin speak the "truth" about the Bolshevik Revolution but Balabanov is subverting it. In *Morphine*, the Revolution is a distant event that makes no sense even to those manning it at the ground level (soldiers, workers). The Jewish intellectual[24] who is a member of the RSDWP is casually shot dead by Polyakov at the conclusion when the latter is escaping from the de-addiction clinic. The intellectual has represented the sense of "socialist purpose" in the narrative and his casual death in such circumstances heightens the irony of the illiterate guards he is leading actually undertaking "revolutionary action" on behalf of the distant Bolshevik state.

Morphine is not ferocious like *Cargo 200* but it is deeply ironic. Its value rests largely in telling a story dealing with a clinical crisis (Polyakov's addiction) and showing that it is unrelated to the historical churning its drama is set in. The personal being political is wisdom in Communism and the early filmmakers of the 1920s interpreted it as members of the working class participating in building history. Balabanov disrupts the notion by showing people too preoccupied with everyday calamities to be even mindful of "unfolding history." Underneath is the sense of human beings essentially physical and irrational forced into a paradigm based on rationalist political principles. I have used the term "people" here but Balabanov's overwhelmingly physical characters are more like awkward bodies coping with an inhospitable world, which is why the motifs of illness and medical procedure are politically effective.

A Stoker (2010)

Like *Dead Man's Bluff*, the film is set in the 1990s and deals with the crime scene in the Yeltsin era. But the film is not a comedy and has the elemental simplicity of a folk song. It could even be compared to Ingmar Bergman's *The Virgin Spring* (1960) in being about a father's revenge for the murder of his only daughter; Bergman's film was also based on a ballad. The protagonist of Balabanov's film Ivan Skryabin (Mikhail Skryabin) is a Yakut and an Afghan war veteran with the rank of Major—and a declared Hero of the Soviet Union—working as a stoker.

His wife has left him and living in Detroit but his only daughter Sasha also lives in St Petersburg, trading in furs with Masha, her Russian partner. Masha's father was a Sergeant and a sniper in the military alongside the stoker but is now a gangster and hitman. Working as his assistant is Vassily or "Bison," a silent man who does his bidding. Bison is lover to both Sasha and Masha although neither knows about it and Sasha hopes to eventually marry him.

Between the moments when he has to feed the furnaces to keep them burning, the stoker (who lives on the premises) is writing a novel set in Yakutsk and about the dealings of the locals with Russian criminals exiled to their land by the former Tsarist state. The Sergeant and Bison use the stoker's furnaces to burn the corpses of those they have hit, explaining to the simple man that the dead people are "bad." The Major tells stories about bad men to a little girl who visits him, the daughter of an air force colonel selling armaments, who has never been to war.

The film uses a narrative with interludes in which people travel to locations where the action happens—an assassin with a guitar case walking to the site of a hit, Sasha trudging to the basement where she works as a furrier, a vehicle loaded with corpses driving to the stoker's furnaces—and Balabanov has instrumental music repeating cyclically as in a jukebox, in the same order in these interludes. Since the action that concludes these transitory interludes is violent crime or implies it, the strategy conveys the sense of brutal crime as a habit in snowbound St Petersburg. The intervals of mechanical sex between Bison and his two lovers complete the picture.

It is in this scenario that Masha discovers that Bison is also a lover to Sasha and persuades her father to have her hit. Bison does so and the corpse is duly burned at one of the stoker's furnaces. The corpse is in a sack and explained away as that of a bad woman and a whore, but a shoe drops from the sack and the Major recognizes it. He goes in search of Sasha and finds the other shoe from the pair in her apartment. He now dresses up in military clothes, complete with his medals, proceeds to the Sergeant's apartment and kills both the Sergeant and Bison with a ski pole. Balabanov changes the music during this last transition on a tram to intense vocal music.

The film moves with the Major back near his furnaces, cutting his own wrists. But the film actually concludes with a frightening black-and-white segment representing a section of the incomplete novel he has been writing, as read out by the little girl visiting him. In that segment, Kostya, a Russian bandit in Yakutsk who has forcibly taken shelter in Habjik's shack, beats him and brutally rapes his wife Keremes, after which she has to endure the violence of her own husband (for having "submitted" to Kostya) that she can only beseech to be made less harsh.

Balabanov uses a laconic filmmaking style to emphasize the irony of Major Skryabin's situation as a loyal "Soviet citizen" after the Soviet Union became "Russia" under Yeltsin; but it emerges as a deeply moving film, made abruptly disturbing by its concluding section. The realistic final section also contrasts with the stylized narration of the overall film, awakening us to a past horror among the gentle Yakuts from Russians—included in the Major's vengeance. We have seen him typing away at his story but never understood what he harbored in his heart against Russians when he had been serving them thanklessly in the army.

Balabanov, as a post-Soviet filmmaker, is not someone once optimistic about a Communist utopia and, in his films too he is primarily dealing with Communism's putrefying legacy. The polyphony in his films is even symptomatic of a disparate population never constituted as a "people." In some films (*Of Freaks and Men*), the failure of the Communist state is ahead while in others (*Dead Man's Bluff*), it is in the past but it looms as the implied subject in all of them, even in *Happy Days* where it is registered as an interval represented by the protagonist's loss of memory. If the reflection on the Communist state is manifested strongest in the films set in the Soviet or post-Soviet periods, even in the films set in prior times, he is doing it by parodic invocation of early cinema, the site that still lends the Communist utopia respect. The pedestal on which the USSR placed cinema after the Revolution makes it Balabanov's ironic reference point.

NOTES

1. The film won the Palme d'Or at Cannes in 1957.
2. Sergei Paradjanov (who was Armenian) was considered a subversive and his bisexuality, especially, made him a target. The Soviet authorities sentenced him to five years in a hard labor camp for "a rape of a Communist Party member, and the propagation of pornography." 'Ukraine Clears Sergei Parajanov of 'Homosexuality' Charges Posthumously', The Armenian Report, 10 January 2024 <https://www.thearmenianreport.com/post/ukraine-clears-sergei-parajanov-of-homosexuality-charges-posthumously> [accessed 19 February 2025].
3. Mikhail Bakhtin notes the conflict between the propagandizing impulse in Tolstoy and the "internal dialogism"; the Marxist polemic of *Mother* has perhaps the same relationship with the polyphony just noted and introduced by the choreography. See Mikhail Bakhtin, "Discourse in the Novel," in *The Dialogic Imagination: Four Essays*, ed. by Michael Holquist, trans. by Caryl Emerson and Michael Holquist (Austin: University of Texas Press, 1981), p. 283.
4. The eyeline match—a person's glance is cut to another's to indicate their respective positions in relation to each other and character glance becomes a clue to link shots—can also be

seen as the acknowledgment by one person of another's existence. In Pudovkin's film, the father is almost oblivious of his family until mother and son accost him physically and he is forced to look. This is not the case in Donskoy's film in which acknowledgment comes first.

5. Gorky regarded his novel as inferior and it was written as a polemical/ideological (and sentimental) counterpoint to Dostoyevsky's portrayal of revolutionaries (*The Possessed*) but it was also considered "timely" by the Communists. Being "timely" meant that it fulfilled a purpose attributed to art/cinema in the Stalinist epoch. While Pudovkin's film was even avant-garde in its aesthetics, Donskoy's film has been described as the only "true" adaptation of Gorky's novel. But coming out after the death of Stalin and in the period of Khruschev's thaw, it was castigated for conforming to "socialist realism"—for which it would have been unequivocally lauded three years before—when Stalin was still alive. See Evgeny Dobrenko, "Three Mothers: Pudovkin-Donskoi-Panfilov," in *Stalinist Cinema and the Production of History: Museum of the Revolution*, ed. by Evgeny Dobrenko, trans. by Sarah Young (Edinburgh: Edinburgh University Press, 2008), pp. 167–79.
6. Mikhail Bakhtin, "Discourse in the Novel," in *The Dialogic Imagination: Four Essays*, ed. by Mikhail Bakhtin, p. 262.
7. In a sci-fi film by Vadim Abdrashitov *Parade of Planets* (1984), one of the focal characters has a conversation with an old woman, the inmate of an institution, who takes him to be her son. He describes his world gloomily in this way, "Everything has been discussed and everything is clear."
8. Nancy Condee, *The Imperial Trace: Recent Russian Cinema* (New York: Oxford University Press, 2009), pp. 217–18.
9. Yana Hashamova, "Aleksei Balabanov's Russian Hero: Fantasies of Wounded National Pride," *The Slavic and East European Journal*, 51.2, Special Forum Issue: "Resent, Reassess, and Reinvent: The Three R's of Post-Soviet Cinema" (Summer, 2007), 295–311.
10. The title invokes the 1961 two-act play by Samuel Beckett but bewildered critics have already noted that the film seems to have little to do with the original other than a stance of grim mockery and a stylized portrayal of disenfranchised life. Hashamova, p. 298.
11. Hashamova, p. 220.
12. Taking off his hat would perhaps be like Popeye shedding his pipe or his sailor's cap.
13. "The generation coming of age after 1991, including those who would discover Balabanov's cinema, entered a time when the encounter with the state would be profoundly different from the experience of their elders. The youth league and Party meetings, state parades, volunteer brigades, official rallies, and demonstrations of the Soviet years had been relegated to historical footage." Nancy Condee, *The Imperial Trace: Recent Russian Cinema*, p. 218.
14. As noted earlier, Balabanov had contributed a segment "Trofilm" (1995) earlier to a compilation film *Arrival of a Train* (1995) to commemorate the Lumiere's film. But the train also played a key role in the Revolution and many filmmakers (like Dziga Vertov) traveled on trains during the Civil War shooting footage, and this became widely celebrated.

15. Peter Daniel DiPaola, "The Criminal Time Bomb: An Examination of the Effect of the Russian Mafiya on the Newly Independent States of the Former Soviet Union," *Indiana Journal of Global Legal Studies*, 4.1 (Fall 1996), 145–81.
16. Dmitry Shlapentokh, "Drunkenness and Anarchy in Russia: A Case of Political Culture," *Russian History*, 18.4 (Winter 1991), 457–500.
17. Stanley Kubrick used classical music in the same ironic way in *A Clockwork Orange* (1971).
18. The first episode in *Pulp Fiction* with Jules and Vincent dealing with the group led by Brett is shocking in its abruptness and, in a sense, puts us in a state of awe with regard to criminal brutality. Balabanov refrains from shocking us in the film and its comedy is not interrupted by shock.
19. Nancy Condee, *The Imperial Trace: Recent Russian Cinema*, p. 231.
20. Birgit Beumers, *A History of Russian Cinema* (New York: Berg, 2009), pp. 78–79.
21. Andropov was a KGB chief but is regarded highly in Russia for going after the corrupt officials of the Brezhnev era but he died abruptly and there were rumors afloat that he was poisoned. Chernenko was a throwback to the Brezhnev era. https://www.britannica.com/place/Soviet-Union/The-Interregnum-Andropov-and-Chernenko. Accessed January 29, 2025.
22. I would propose that terror is abject fear and the opposite emotion from desire, while horror is the opposite of wonder. The two are therefore different emotions and horror films do not always evoke fear.
23. Among others filmmakers to place emphasis on the body as the seat of the "person" are David Cronenberg (*The Fly*), Patrice Chereau (*Queen Margot*), and Bruno Dumont (*Humanity*). The tendency is to contrast the body with the person and in *The Fly*, as the body alters so does the person's nature.
24. His name is Grinburg; Leon Trotsky (Lev Bronstein) was also Jewish and a Bolshevik leader. Grinburg is made to look a bit like Trotsky.

8

David Lynch: Public Mythologies and Personal Fantasies

David Lynch is perhaps the most mysterious of all American filmmakers, a factor made more striking by the fact that his films are eminently watchable and also have a large following. He is usually described as "surrealist" and, as in Luis Buñuel's case, I detect a strong satirical impulse in his films although large parts of them might not readily answer to the description of the term, since the social element in them is not easy to identify. He has called himself "apolitical" and it is not unlikely that he relies more on his sub-conscious self than Luis Buñuel, who was consciously political even in his Mexican melodramas. It is difficult to classify Lynch's films separately into sections but given that his work in television—notably his series *Twin Peaks*—is important, I will treat each of his films separately and then devote a separate section to *Twin Peaks*, or at least to the last part entirely created by him. His avant-garde/independent feature *Eraserhead* may give us some insights into his approach since it is one with his later work. Critics are uncomfortable interpreting it in the accepted sense, but not doing that would amount to treating it as mere sensation. Hopefully, it will give us an opening into the project of interpreting the body of his work, culminating in *Twin Peaks: The Return* (2018), one of the most successful efforts ever at raising a television series to the level of high art. Lynch is a mainstream filmmaker and I will treat him as that, using the *Eraserhead* as an entrance point, then going on to his "mainstream" films. Needless to add, I will only examine some of Lynch's films, the purpose merely being to get a grip on his methods.

Eraserhead *(1977)*

In the film, a man Henry Spencer (Jack Nance) lives in a seedy one-room apartment located in a nondescript industrial wasteland. One day he gets a call from a girlfriend Mary X (Charlotte Stewart) he has not seen in some time. She has given birth to a baby which is apparently premature. Her mother insists on their marriage. In the next segment, he and Mary are together looking after the child, a grotesque rabbit-like creature wailing much of the time and, except for its shiny hairless face, swathed in bandages. Unable to cope with the baby, Mary leaves. Henry looks after it through an illness when its face erupts in tiny boils and it refuses to eat, but he tries to care about it. He has a brief sexual encounter with a neighboring woman, perhaps a prostitute. Insistent visions of a strange woman who exists behind the radiator and performs on the stage singing innocent songs impinge on his claustrophobic world. Finally, he cuts the bandages wrapping the baby with a pair of scissors and discovers that the bandages have held its vital organs together. He cannot reassemble the creature and he stabs one of its organs with the scissors and kills it, also seeing it flooded with a light-colored excretion.

Eraserhead, writer-director David Lynch's first feature, has been baffling and disturbing audiences since its release. It has become one of the most persistent and successful cult films on the midnight and art-house circuits. Such films invite the audience to throw off their inhibitions and join in an assault on generally accepted standards of good taste. But *Eraserhead*, while it shocks and seems even perverse, is too introspective and does not allow young people to unite against a common hatred like "bourgeois attitudes" or the "political system":

> It provides an auditory and visual assault which isolates each viewer. The experience becomes intensely personal, unshared. Lynch achieves this by relentlessly applying alienating devices. Foremost among these is the setting of the film in a bleak world not recognizable as our own. The action which takes place there offers no narrative with an externally meaningful coherence. The simple "story" of Henry Spencer's forced marriage to Mary X is utterly banal—yet it is couched in a collection of bizarre, seemingly meaningless images and inconclusive scenes which shatter the story's familiarity and make it frighteningly strange.[1]

Eraserhead has such an impact upon our senses that it would be helpful to describe it before trying to extract some meaning out of it. Henry is a printer and the world he lives in, while not our own, can still be recognized as an expressionist version of it, and Mary's family could be inhabiting the kind of home lived in by Andrei Tarkovsky's *Stalker* (1979), set in a nightmarish urban/industrial world, the noise of a passing train periodically drowning out household sounds.

The family has the father, the mother, and Mary and all of them are afflicted in some way, the father (a plumber) describing the numbness in his arms to entertain Henry, while both Mary and her mother have seizures, although of different kinds, evidently a familiar occurrence. The chicken served is reminiscent of the dishes conjured up by chief inspector Oxford's wife in Hitchcock's *Frenzy* (1972) which he can barely keep down; in the film it twitches on Henry's plate. The chicken is fist-sized, lets out a dark stream, and evokes revulsion in Henry, who is asked to carve it. When Mary's mother takes him aside to mention the baby (Mary is not sure it is a "baby"), she makes sexual overtures to him and Mary, who has apparently also seen this before, has to pull her way. Henry then has a nosebleed, which has Mary exclaiming anxiously.

The existence described by Lynch is banal but the portrayal he offers is hardly that; *Eraserhead* is the banality of domesticity made horrific. The film opened with a nightmare of a "spermatozoa" (turned into a squelchy reptilian creature) coming out of Henry's mouth and fertilizing the earth, as it were, and in Mary's house is a dog attending to a litter of over half a dozen puppies, and this aspect of the film can be understood as having to do with fertility, but with horror instead of in the familiar way. The "Man on the Planet" who seems to have a hand in all these matters, a sick old man who works on a handful of levers, is the occult element underlying life, interpretable as God. The Man on the Planet seems to be sick in the same way that the baby is sick.

All these aspects of *Eraserhead* dealing with domesticity and the nuclear family can be described as ironic because it takes an antagonistic position of the customary valorization of the notions by Hollywood. If, in the light of Althusser's notions of the family as ISA (Ideological State Apparatus),[2] Hollywood's treatment of the nuclear family is regarded as an affirmation of the dominant ideology of the nation-state, Lynch's film is politically subversive.

This brings to other aspects of the film, of which the chief one is the girl in the radiator who sings an innocent song ("In heaven, everything is fine"), squelchy spermatozoa raining around her even as she does it. My proposition is that the radiator is only a conjunctive device intended to bring her into the film since Lynch frequently uses interludes of popular music (as performance) in his films and they seem to provide moments of tranquility amidst the horror and anxiety pictured, and that could be true of this girl as well, not as a plot device but indicating Henry's desire for normalcy in his life. That the girl has her face made up grotesquely while making her seem "innocent" is to keep her aligned with the bizarre tone of the film. Her stage performance contaminated by squelching spermatozoa is simply Henry's nightmare contaminating his searching after peace. Henry's is decapitated after this in the "heaven" on the stage and his head replaced by the baby's; his head, deposited on the street, is lifted by a passing boy, found to be of rubber, and then used to make erasers in a factory. One cannot be sure what to make of all this but surrealism is habitually making bizarre and incongruous associations. The eraser could however be taken to be a wish for "erasing the error" Henry made with Mary. We recollect that the spermatozoa originally came out of Henry's head. I am not proposing a one-for-one correlation but only a broad basis for interpretation.

I will also try to demonstrate that each Lynch film should be taken in its totality and not as a collection of interpretable elements.

If one were to begin to "understand" *Eraserhead*, it must be described as a very personal film with Henry perhaps representing Lynch himself. Henry is perpetually in a crumpled suit and tie, with a mass of unruly hair on his head, and that is how Lynch usually presents himself in public. The film is, broadly speaking, about the fears of fatherhood and domesticity and the protagonist's unsuccessful efforts to be rid of these anxieties. The film is personal but with one qualification, which is that Henry is presented as a social animal and not an embodiment of "feelings"; we do not see the world through Henry's eyes and he too is caricatured and mocked. Lynch is discounting personal feelings and presenting Henry's existence as a human condition, although in a personal way. In the process he shows us the "world," "heaven," "God," "creation," "life," and "family" in his film and these are all notions very important in the social milieu, but *as political emblems*. "Life," for instance, is a notion that dominates political debates at election time.[3] The killing of the malformed child by Henry, if understood for what it was, could create great unease within a vocal political segment in the United States. By giving manifestation to all these emblems in such a perverse way, Lynch is being deeply sardonic[4] although, as with his other films, we cannot know how deliberately that is intended.

Many of the images in *Eraserhead* are to be repeated in his later films but with the same result, which is why we cannot interpret them individually but taken together they convey something. Squelchy creatures, dark liquids oozing out of solid bodies and dry, leafless bushes, for instance, appear and reappear in his films, and these are only a few. While Lynch may be working extensively with his irrational side, appreciating his work can only be done by looking at its totality in rational terms, or at least large parts of it. I also surmise that Lacanian psychoanalysis, often used on Lynch,[5] may not be the only way to approach him although it has its own legitimacy since there is the personal side to his films. But here is an observation made about Lynch's films that I need to take issue with since, if it is completely true, reading his films as political satire might not be feasible:

> Lynch's films explore the psyche to such an extent that they never seem to touch the ground, to engage the economic and political realities that shape our everyday lives. If there has been one sustained theme of criticism of Lynch's work, it has followed these lines: he creates filmic worlds that show little sign of the material world- of class inequality, marginalised people or economic struggle.[6]

It is true that there is no portrayal of class inequality or economic struggle in *Eraserhead* but my argument was that society is present in its narrative and it tries to implicate key dominant discourses in the public space, here especially the sacred notion of having children and caring for them; there is a covert introduction of religious motifs through "the man on the planet" who, I suggested, represents God, and is blamed for Mary's pregnancy and the malformed child. Henry copes with his situation through infanticide and, in the society that Hollywood represents through its narratives that is a grave sin—which makes the motif a deeply divisive political reality.

Henry may be trying to escape from his social reality through his fantasies[7] but at the climactic moment in the film, Henry is engaging with his reality by killing his own child. The film may be "personal"—not least because Henry is apparently Lynch himself—but Lynch, even when he represents his own experiences, is seeing himself as socially involved, with public mythmaking about the family—rather than economic relationships—as his primary preoccupation. The exaggerated histrionics that Lynch makes the actors engage in is also distancing. Lynch is representing the characters in *Eraserhead* largely as caricatures which also suggest satirical purpose. It is this social aspect of his films that justifies my reading them as irony and satire; psychoanalysis, even if legitimate in a different context, can be arguably put aside for my purpose. It may throw light on Lynch's "person" but my interest is elsewhere. *Eraserhead*'s horror of domesticity may have been "personal" since Lynch had become a father at only 22 but it still has socio-political implications.

Mainstream films

Eraserhead established Lynch as a new voice in independent cinema, but it screened primarily at midnight showings in New York City, which allowed it to attract a cult following, directors like John Waters and Stanley Kubrick noticing and appreciating it, but it was Mel Brooks's love of the film that transformed Lynch's filmmaking career. After a private screening, Brooks offered Lynch the chance to direct a project he was producing for Paramount. This project became *The Elephant Man* (1980), the film that brought Lynch public and critical recognition,[8] which might have been elusive otherwise. The easiest way to deal with it is to treat it as a career move that simply demonstrated Lynch's competence at conventional filmmaking, but one might still find something apart from the humanist project it was purported to be.

The film is based on a true happening of Victorian England and Frederick Treves (Anthony Hopkins), a surgeon at the London Hospital, finds John Merrick (John Hurt) in a freak show in London's East End, where he is kept by Mr Bytes, a sadistic, and heartless showman. His head is kept hooded, and his "owner," who views him as intellectually feeble, is paid by Treves to bring him to the hospital for examination. Treves presents Merrick to his colleagues and highlights his monstrous skull, which forces him to sleep with his head on his knees, since if he were to lie down, he would suffocate. On Merrick's return, he is thrashed so badly by Bytes that he has to call Treves for medical help and Treves brings him back to the hospital. Merrick is tended to by Mrs Mothershead, the forbidding matron, who voices many of the moral concerns needing to be expressed. Mr Carr Gomm (John Gielgud), the hospital's Governor, is against housing Merrick, as the hospital does not accept "incurables." To prove that Merrick can make progress, Treves trains him to say a few conversational sentences. Carr Gomm sees through this ruse, but as he is leaving, Merrick begins to recite the 23rd Psalm, and continues past the part of the Psalm that Treves taught him. Merrick tells the doctors that he knows how to read, and has memorized the 23rd Psalm because it is his favorite. Merrick is literate and intelligent and not the "imbecile" Treves took him to be, given his habitual silence. Carr Gomm permits him to stay, and Merrick spends his time practicing conversation with Treves and building a model of a cathedral he sees from his window.

Lynch takes up a subject that virtually dictates how it should be treated—in an "uplifting" or "inspirational" way and affirming humanity. The way that would be done by a director would be to show the victim bearing his cross bravely, and a film that might bear comparison is Peter Bogdanovich's *Mask* (1985) the true story of a teenager suffering from "craniodiaphyseal dysplasia," causing severe skull deformities. Empathy is a path Lynch does not obviously take and John Merrick is not someone we are made to identify with, which is the way Bogdanovich goes about it. Of course, Merrick is a more grotesque figure and such identification would be difficult, but why is so much attention given to Merrick's exhibition at various levels? The sequences in which people of different stations visit him seem repetitive and manipulative, until we notice that these viewings by people are hierarchically categorized, class-wise, and often acted out as hyperbole. The lower order is distinctly threatening while the attention of the higher order is as if to be desired. When Mrs Mothershead points out that they are not really different Treves is disturbed, but that line is not pursued since it would interfere with the film's "inspirational" content, which is why it became a project in the first place.

The recurring vision of elephants associated with Merrick and his "angelically beautiful" mother may also correspond to an occult or religious element that Lynch often introduces to raise the level of the narrative from being overly mechanistic, dwelling only on Merrick, and we see this occult element in his later films as well. In a sense, Merrick is blessed with a "sign"[9] more than cursed and a key scene in which he is treated with reverence by other fairground "exhibits" (a giant, dwarfs, Siamese twins) and put on board a ship for England demonstrates it also. If one wonders who his appearance is a "sign" to, it could be this community of "freaks." *The Elephant Man* is a more complex work than may seem at first glance and while one cannot see it keep the company of Lynch's other films, it is more than morally uplifting.[10]

Blue Velvet (1986)

Lynch was born in a small-town Missoula, Montana, and lived his early life in other small towns. That had an impact upon him and the motif of the small town, which finds a place in *Blue Velvet* is also in *Twin Peaks*. The factor that we associate with Lynch's view of the small town are the dark goings on underneath, crime and perversion, belying the outward look of normalcy.

The film begins in bright sunshine in an idyllic suburban town named Lumberton—with Bobby Vinton' song "Blue Velvet" heard on the soundtrack. Children are crossing the road on their way to school, a man on a fire engine is waving, and the local radio station is making cheerful announcements. In a garden an elderly man is watering the lawn until matters abruptly get out of hand: the pipe becomes twisted around a bush, the water gushes out at the wrong end, the dog is excited by it, and the man has stroke, collapses on the grass, the water spraying in all directions. The camera now gets under the sunny piece of lawn to look at insects, beetles, and worms as the music suddenly turns sinister. In the next scene, Jeffrey Beaumont (Kyle MacLachlan) is visiting his father in hospital. He takes a shortcut across a field and on his way back by the same route discovers a severed human ear covered with ants. He takes it to Detective Williams who is cool but interested and promises to pursue the case. Going out, Jeffrey meets Sandy Williams (Laura Dern), the detective's daughter who has heard him and tells him that her father is investigating a case connected with a night-club singer named Dorothy Vallens (Isabella Rossellini) and that this strange find may have something to do with it. The two survey her apartment block and, later, visit the lounge in which she is performing and listen to her singing "Blue Velvet." Searching her apartment is too tempting a prospect to pass up and Jeffrey pretends to be a bug exterminator, visits her at home, and finds a way to get back inside when she is away singing and come away with a key to facilitate it.

A small town has places in it with histories or mythologies—like dilapidated buildings or residences inhabited by strange people; films like *To Kill a Mockingbird* (1962) invoke such places as part of their design. The heard stories may not be true but they are usually dark and the same is true of ordinary people in such places whose appearances attract curiosity when one does not know much about them. Where, in a city, the sensations one receives are too many and overwhelming for any to remain etched in one's mind, the small town is the natural breeding ground for bizarre stories—given substance by the restrictions placed on activity and the impossibility of anonymity. Here is Lynch on his childhood in the small towns he grew up in:

> My childhood was elegant homes, tree-lined streets, the milkman, building backyard forts, droning airplanes, blue skies, picket fences, green grass, cherry trees. Middle America as it's supposed to be. But on the cherry tree there's this pitch oozing out—some black, some yellow, and millions of red ants crawling all over it. I discovered that if one looks a little closer at this beautiful world, there are *always* red ants underneath. Because I grew up in a perfect world, other things were a contrast [...] I found the world completely and totally fantastic as a child.[11]

The key aspect of Lumberton in *Blue Velvet* is that it is banal and the conversation is constructed to emphasize this. When Jeffrey, whose father owns a hardware store, wants to investigate Dorothy Vallens by going into her apartment alone, what Sandy asks is whether he is a detective or a pervert, the two options open to anyone who might want to get involved in a local mystery of this kind. The rest of *Blue Velvet* can be described as nightmarish since Jeffrey discovers that Dorothy's husband and child have been kidnapped by a man named Frank Booth (Dennis Hopper) to force her into sexual perversions of various sorts. Booth has several associates who are engaged in some kind of vaguely criminal activity—perhaps involving drugs. Many of these villains are identified almost in comic book fashion—a suitcase, a yellow suit.

Detective Williams always seems to know what Jeffrey reports but he never says more than "we have been watching them" or "they are very dangerous people." In the midst of all this, Jeffrey becomes involved with Dorothy Vallens though the relationship is violently obsessive; it is as if Jeffrey has been *aroused by her predicament*. It also turns out that the ear discovered by Jeffrey belongs to Dorothy's kidnapped husband and was cut off to pressurize her. The film extracts emotions from us that are affecting but often contradictory. Jeffrey sees Frank Booth snarling, violent and foul mouthed and trying to get Dorothy to acquiesce to perversions but a short while later Jeffrey and Sandy see him at the nightclub listening to Dorothy singing "Blue Velvet," tears streaming down his face as if deeply affected in emotional terms.

When one were to look at the plot of the film, especially its "dangerous villains," one might find it even laughable; the evidence Jeffrey presents to Detective Williams are innocuous pictures of people talking to each other in some corner. A drug dealer is killed in the vicinity of where the two people are talking and this is mentioned in hushed whispers. The word "drugs" is used simply to convey the sense of something irredeemably evil and perverse. Frank Booth is a "dangerous man" but more like a caricature of such a person in a child's mind. To provide the reader an idea, one of his fiercest diatribes is against Heineken (he prefers Pabst Blue Ribbon), while another furious one is against beer being warm! His associate Ben is effete and lip-syncs a pop song, Roy Orbison's "In Dreams," which disturbs Frank, who goes off into a sulk. But his strongest association is with sexual perversions and, very soon, Jeffrey himself is also in an intense sadomasochistic relationship with Dorothy, who insists that he beat her.

Blue Velvet, as has been noted by several writers, is neatly divided between the banality and ordinariness of Lumberton—an idealized American fantasy—and what might be termed its "evil double," which is a darker, perverse fantasy about the underside of the same denoted small-town space:

> The public world that we see when the film opens is not "the real world" but a purely fantasmatic one that corresponds perfectly-even too perfectly-to an American ideal. The opening [...] images suggest the kind of perfectly realized fantasy world that one never encounters in rea life. On the other hand, the horrific underside of this public fantasy is equally extreme. Frank Booth and his gang revel in their perversity and total disregard for the public law. Mere exposure to them endangers the life of Jeffrey Beaumont [...] and results in a severe beating. If the public world of *Blue Velvet* represents an American ideal, its underside represents an American nightmare. What distinguishes *Blue Velvet* from the typical American fantasy is the way it holds these two worlds apart.[12]

The solution offered to this division is characteristically Lacanian, but such decipherments are of a general kind pertaining to the divided nature of fantasy.[13] Still, the film is even comic in its hyperbole and I will, to interpret it, describe a sequence involving Sandy and Jeffrey. After going to attend a party, their car is chased furiously on one of Lumberton's empty streets and, since we have already seen Frank Booth racing similarly, we (and Jeffrey) take it to be him. What happens instead is that in the car is Mike, Sandy's jealous boyfriend, who has noted Jeffrey's growing intimacy with Sandy and has come to "kill" him. Both cars stop in front of Jeffrey's home but before Mike can "kick his ass" as he intended Dorothy appears naked in Jeffrey's porch, battered and bruised. In a few moments, Mike is apologizing profusely to Jeffrey, although a friend reminds him of his purpose: "Man, I thought you were gonna kill him."

There is mimicry in this segment by Mike's friends of Frank's gang but a distinction is made and the issue is why the naked Dorothy should frighten the boys into fleeing. They know nothing about her but it is as though the threat they hold is pitiful compared to what she represents. Mike, as a bully, is part of small-town unpleasantness from youth films but he is in awe of the naked Dorothy and an analogy would be the way hoodlums might be if confronted by a ghost, not a private hallucination but something actually manifest. This leads one to suspect that the two "imaginings"—the small-town film ideal and its dark, supposed opposite—exist at different levels; they do not fit neatly as antitheses of each other since they are not both "public fantasies."

The small-town ideal, to elaborate, is part of a national mythology reflected in mainstream films from Frank Capra's *It's a Wonderful Life* (1946) to Jim Jarmusch's *Paterson* (2016), while the dark one is specifically "Lynchian fantasy," with even self-parody in evidence. We respond to Frank Booth not as a social (or even generic) sign[14] but only as a Lynchian picture of perversity. It is so perverse that mere social wickedness blanches at its sight. The analogy of the "ghost" is apt because Lynch is frequently dealing with the occult and much in the film defeats explanation—like Dorothy saying about Jeffrey, "He put his disease in me." But ultimately, we laugh because of the way Lynch demolishes a public myth familiar to us through genre representations by opposing it with an incongruous creation that comes entirely out of his psyche.

It has been noted that the film depicts two different worlds, the one that we experience as the real world that we see, hear and touch, and a dream world which must remain hidden, so potentially dark and violent are its wanderings.[15] But this can be contested since Lumberton on the surface is not the "real" small town but the way it is constructed in generic exercises and part of public mythmaking about community life. A grittier picture of the small-town life might be found in Peter Bogdanovich's *The Last Picture Show* (1971), rather than the films of the genre that Lynch is (arguably) lampooning. Frank Booth's fearsome "hatred of warm beer" is a clue that Lynch is not taking his own picture of evil as seriously as critics take it.

Lynch is reported to have said that *Blue Velvet* is about "what is hidden" but I take that to include personal imaginings as well, although not "public myths" which are far from hidden. But Lynch mixes up reality, social mythology, and his own kind of partly occult visualizations; the Lynchian fantasy debunks socio-political mythmaking—here the mythology around the idyllic small town—through a hint of the horrible unknown that social mythologies want no part of. The Lynchian fantasy is like the ghastly artistic visions[16] placed in opposition to the certainties or beliefs of their times, and the myth of the small town perhaps had a counterpart in the reassurances of religious myths. The transgressions, the voyeurism, the violence, and the perversion, testify to that purpose since they are placed in opposition to the certainties of its own time, through cultivated mythologies. Lynch's dark private fantasies are oppositional *and anarchic* responses to the *social mythologies* furthered by film genres and are best understood by making that connection. The nasty humor present in the "dark fantasies" justifies the appellation "satirical" applied to him. To make another link with the "personal" element in satire, *Un Chien Andalou* used the dreams of its creators but debunked social mores. Lynch's "surreal" characteristic lies similarly in his fantasies being effectively adversarial.

Wild at Heart (1990)

There is much in *Blue Velvet*, especially around its dark side, that cannot be interpreted by the audiences responding to it and that would include its principal villain Frank Booth and what he represents. Frank Booth is so excessive and constituted by so much capricious detail—like the gas mask that he pulls out and inhales something through—that the audience could be responding to him purely as sensation—made captivating by Lynch's mise en scene. I interpreted the film by examining this "dark" section only in relation to the conventional parts, how their juxtaposition worked and what it might actually mean. *Wild at Heart* is, to a much greater degree, constituted by excess, and it is difficult to find a reference point in relation to which it might be interpreted. To elaborate, if cinema has a real and another expressive component, the relationship between the two is the basis for the kind of interpretations I have been favoring, but the "real" side of *Wild at Heart* is even smaller than that of *Blue Velvet*.

Wild at Heart is a road movie, a genre already touched upon in the chapter about Aki Kaurismaki, and road movies deal with the random journeys of individuals through a space, the people they meet, and the experiences that befall them. They are also about people who are without big ambitions, people who are usually drifters, and often on the run, pursued by other people. By and large, their journey is intended to lead to an understanding of a designated space, finding unity in it or determining its essence, and since that space will need to have an identity for a unity to be made of it; it is usually a single country that they travel though. Rarely, for instance, do the characters of a road movie cross a national boundary. Films like Ridley Scott's *Thelma and Louise* (1991), Peter Fonda's *Easy Rider* (1969), and Wim Wenders's *Kings of the Road* (1996) are all road movies that answer to this description, something that Kaurismaki subverts in *Leningrad Cowboys Go America* when the band leaves America behind.

In Lynch's film lovers Lula (Laura Dern) and Sailor (Nicholas Cage) are separated after he is jailed for killing a man who attacked him with a knife; the assailant, Bobby Ray Lemon, was hired by Lula's mother Marietta Fortune (Dianne Ladd) to kill Sailor, but Sailor beat Bobby's head against the stone floor till it split and his brains spilled out. In killing Bobby Ray Lemon, Sailor goes much beyond the justification of self-defense and it is as though he is stamping his personality through the act. Upon Sailor's release, Lula picks him up at the prison where she hands him his snakeskin jacket; Sailor declares that it is a "symbol of his individuality and sense of personal freedom." Sailor and Lula are sexually besotted with each other and after checking into a hotel and making love they go to a club where Sailor does a rendition of an Elvis song "Love me." Lula asks him why he did not sing "Love me tender" instead and Sailor replies that he will only sing it for the special woman who will one day become his wife.

The motif of lovers chased by potential assailants is a familiar one and *Wild at Heart* alternates between Sailor and Lula and Marietta's machinations to have Sailor killed. Marietta first sets private detective Johnnie Farragut (Harry Dean Stanton) her off-on lover after them and he is the only person to point out that Sailor is not a murderer but acted in self-defense. But Marietta, unsure of Johnnie Farragut, also hires a gangster Marcello Santos to kill Sailor and the latter kills Farragut instead, after torturing him. We gradually come to know many things about Marietta, primarily that she had her husband Clyde burned to death by Santos and that Lula's uncle sexually abused Lula at the age of thirteen, after which the uncle was also murdered. Sailor, who was driver to Marcello Santos, was witness to it, and therefore the need to kill him.

There is a huge amount of violence but none of it seems to matter since it is all done comic book style. Sailor and Lula are, meanwhile, driving through Texas, he breaking his parole to do so. In Big Tuna, Texas, identifiable by a sign "Fuck You," they meet Bobby Peru (Willem Dafoe with bad teeth) who, after forcing himself upon Lula, touching her sexually and making her say, "Fuck me," involves Sailor in a robbery and has his own head accidentally blown off; Sailor gets five years, but Bobby Peru was going to kill him. Lula has been pregnant—their room in Big Tuna smelled of vomit—and their son Pace is born when Sailor is in jail. Sailor comes out feeling he has to move away from Lula and their son; but a good witch (from the *Wizard of Oz*) appears in the sky and advises him otherwise. He therefore chases after Lula, and sings Elvis's "Love me tender" to her standing on the hood of the car; they will now be married.

There is much bizarre passing detail—seemingly gratuitous—that plays no part in the outcome; one is a traffic accident that Sailor and Lula come upon, with two dead men and an injured woman, who asks for her lipstick before she dies. Strange people with even stranger names (like "Bozie Spool") walk in and out of the story contributing a warped sense of the people littering the landscape. Bobby Peru is a former marine involved in a massacre in Vietnam. The sense to be got is that replacing one bizarre or violent detail with another would scarcely affect the film's design and that only the totality of it matters. The way to approach the film is apparently not through any of this anarchic detail but through Sailor and Lula, who at least have genre antecedents to compare with. But unlike in most other "lovers-on-the-run" road films (like Robert Altman's *Thieves Like Us*) where the lovers are criminals and killers, the lovers here have done no real wrong and the film has therefore a happy ending. Marietta Fortune and Marcello Santos also disappear from the story, though inconclusively.

In the genre of the road film, of which the "lovers-on-the-run" can be considered as sub-category, the experiences encountered along the way are linked to each other so they come together as pertaining to a single entity, usually the nation. In Wenders's *Kings of the Road* the experiences reflect the dark past of the nation and military defeat and in *Thelma and Louise* the two women are oppressed by patriarchy in various forms, Texas representing an extreme case they are frightened of. When lovers-as-criminals are pursued, there is the sense of they being in danger from their pursuers and it is the danger from without that cements their relationship. Important here is the notion of "individuality" and the protagonists in *Thelma and Louise*, *Bonnie and Clyde*, and Kit in *Badlands* all have conspicuous "style"; the tension in the films is between their individuality and the oppressiveness of the milieu that seeks to erase it.

The individuality trying to stand out against the homogenizing milieu is not in evidence in *Kings of the Road* and the motif is arguably a product of the American cultural system, where the notion is ideologically entrenched.[17] Sailor in *Wild at Heart*, with his snakeskin jacket as an expression of his "individuality and personal freedom" and his actions—from being a potential rock-and-roll star to the way in which he expresses himself by smashing Bobby Ray Lemon's skull—parody the mythology around the notion. It is in this context that the crowding of especially bizarre events and people around the lovers (the "Lynchian fantastic") should be seen—as constituting a surreal setting with which Sailor's extravagant expression of individuality is at odds.[18]

Individuality stands out in the midst of the ordinary and not in an assortment of colorful eccentricities. The ready way in which Sailor and Lula's adversaries evaporate underscores the same thing since it is the strength of adversity that gives individual resolve any value. Lynch's choice of Nicholas Cage in Sailor's role is perfect; Cage is a star but with the rare capacity to play the "individual-as-star" in an especially inflated or parodic way. It is the same with Laura Dern, who plays the glamorous blond lover ("Gentlemen prefer blondes," says Sailor) in an overblown way, even as the smell of vomit fills their room. What Lynch is parodying is a cherished myth in the milieu, which means that it is best to do it equivocally. It was the same with *Eraserhead* where recognition of what he was doing might have caused immense disquiet.

Blue Velvet and *Wild at Heart* use the same strategy which is to invoke generic representations from Hollywood (small town film, road movie) and insert willfully bizarre elements into the narratives that include strange characters given to perversities, though even these perversities are not horrific as much as caricatures of evil—"drugs" or "sexual gratification" used as shorthand as in *Blue Velvet*—and one could compare the comic extravagance of Frank Booth to the Joker's in Tim Burton's *Batman* (1989) instead of the noir villain he passes for. I read the oppositional relationship between genre mythology and personal fantasy as a satirical turn, evidenced by the *comic aspects* of the fantasy side, which is not to assert that the fantastic aspect cannot be interpreted psychoanalytically to get into David Lynch's psyche.

Lost Highway (1997)

David Lynch was described by Pauline Kael as the "first popular surrealist"[19] but his popularity may not have much to do with being a surrealist. As I have argued hitherto, there is a strong social content in his films, largely a debunking of conventions that we also saw in Luis Buñuel. Lynch focuses on generic rather than social conventions but generic conventions point to aspects of the dominant mythology, which we discovered in three out of the four films examined hitherto. His popularity, on the other hand, is more due to the puzzles he sets, especially his kind of narration in which much is held back and the process of narrativity[20] initiated in the mind of the active spectator. His popularity reached its peak with *Mulholland Drive* (2001) and his fans began to regard his films as puzzles to be solved through the clues he provides, rather than subversive in any sense, which is where his value as an artist arguably lies. This is perhaps most true of *Lost Highway*, which does not mean that it has not been seen in Lacanian terms, the favored academic way of dealing with his work.

In *Lost Highway*, Fred Madison, a Los Angeles saxophonist receives a message on his house intercom: "Dick Laurent is dead." The next morning, his wife Renee finds a VHS tape on their front steps containing a video of their house from the outside. As the days pass, more tapes arrive, showing shots of them asleep in their bed. Fred and Renee call the police since it implies that their house has been broken into but the detectives offer no assistance. Fred and Renee attend a party being thrown by her friend Andy. The Mystery Man Fred dreamed about earlier approaches Fred, claiming to have met him before. The man then says he is at Fred's house at that very moment and answers the house phone when Fred calls him. Fred learns from Andy that the man is a "friend of Dick Laurent's." Frightened, Fred leaves the party with Renee. The next morning, another tape arrives and Fred watches it alone. To his horror, it shows him standing over Renee's dismembered body. Renee has been murdered and Fred is sentenced to death for her murder. On death row, Fred is plagued by headaches, visions of the Mystery Man, and a burning cabin in the desert. But during a casual check, the prison guard finds that the man in Fred's cell is not Fred but another person not seen before, who turns out to be Pete Dayton, a young automobile mechanic. Since there is nothing else to be done, Pete is released into the care of his parents, but under surveillance by two detectives, trying to find out more.

The favored way of reading this film is to see its strange twist as imagined by Fred and Fred has had sexual inadequacies as well as suspicions about Renee's fidelity, and has seen her with Andy.[21] But it would still have to be grounded in some reality (or myth) that would also have to be identified. The problem with the Pete Dayton part of the story as imagined is that it contains too much social detail; it is plausible that Fred dreamed up Pete but not the kind of detail put into Pete's story. Mr Eddy, a gangster, is fond of Pete and takes him for a ride in his car and another driver tailgates Mr Eddy (drives behind him, leaving insufficient distance to stop without causing a collision) and the latter is so infuriated that he chases and grievously assaults him. Mr Eddy and his henchmen (on their own) represent a satirical portrait and one cannot accept that he is only the imaginings of someone on death row, wanting to escape from himself. Why, also, should Fred dream up the two policemen keeping tabs on Pete Dayton?

Pete Dayton is a mechanic with the required skills and full memory of his own past but few discernible links to Fred Madison. Mr Eddy, a few days later, arrives at the garage with a blond woman Alice Wakefield, a dead ringer for Renee, with whom Pete commences a relationship. We recognize her as Patricia Arquette who played Renee but little suggests that Pete Dayton "recognizes" her. When Mr Eddy finds out about Alice and Pete, the two try to flee but there are other twists. Mr Eddy is also Dick Laurent and the Mystery Man, who now appears, kills him. Alice disappears, Pete becomes Fred and the film ends with Fred suddenly convulsing and screaming while driving his car, which I take to be his throes on the electric chair. It could be that the film is replicating the strategy of Robert Enrico's *An Incident at Owl Creek Bridge* (1962) in which half of the film is the story running in a man's head at the moment of his death by hanging.

As already indicated, the film does not work well as a plotting puzzle to be reconstituted and, given Lynch's earlier films, I am not convinced that creating such a puzzle is even his purpose. At the very least, I would propose that Fred Madison and Pete Dayton are equally real people. Pete, for instance, cannot remember how he got into the jail either, implying symmetry not consistent with Pete being only dreamed and Fred being the real one. His life is also transformed by the experience and not only Fred's. *Lost Highway* is described as neo-noir, and marital jealousy and murder are recognizable motifs in the genre. Perhaps there is something in the noir genre that would explain *Lost Highway* to us more clearly in the terms I have chosen.

The story of Fred Madison and Renee has the makings of noir but *Lost Highway* may be problematizing the notions of responsibility, guilt, and remorse, and the associated issues of identity and memory of the crime, so crucial to the genre. If a person does not recollect his crime, can he be held guilty of it and accept responsibility is a question it could be asking. Early in the film, a policeman asks Fred Madison if he has a video in the house. Here is the exchange:

COP: Do you own a video camera?
RENEE: No. Fred hates them.
FRED: I like to remember things my own way.
COP: What do you mean by that?
FRED: How I remembered them. Not necessarily the way they happened.

Genre films involving crime presume moral continuity in the individual but what if that was questionable—in this instance if a gap existed between something that happened and the event as remembered? Fred may be punished at the conclusion but can he be held morally guilty? This, it can be argued, is the crux of *Lost Highway* and I take the title to pertain to Fred slipping off the "highway," without the moral continuity in the individual that noir fiction takes as a given. The various twists and turns—including the Mystery Man—are only attendant to making the demonstration and introducing the occult element frequently used by Lynch to subvert dominant mythologies. This is akin to a thought experiment, a philosophically valid way of questioning a dominant supposition, and an instance would be "Maxwell's Demon"[22] in physics. In a thought experiment, a farfetched but logical piece of fiction is constructed to falsify an existing hypothesis.

Examining the mythologies Lynch is interrogating, they are increasingly more elusive. In *Wild at Heart*, it was the cherished notion of individuality and in *Lost Highway*, it is the concept of the stable individual responsible for his or her moral choices. Buñuel posed a similar question in *The Diary of a Chambermaid* but more as a critique of the moral order in Catholic society. *Lost Highway* is, instead, like the subversion of a convention in noir films, viz. that of retribution for wrongdoings. Punishment is rational only if the individual comprehends the reasons for it, but Fred Madison does not. The convoluted scenario chosen by Lynch is arguably only to problematize "moral continuity" by which the criminal act and culpability are part of the same continuum. As in *Wild at Heart*, the cherished notion of the individual is at the center of Lynch's subversion but the difference is the genre vehicle employed each time—the lovers-on-the-run road movie in *Wild at Heart* and noir in *Lost Highway*.

Mulholland Drive (2001)

This film is also generally read as a puzzle to be untangled but the first part nonetheless emerges as a satire on Hollywood. The title sequence features a dance contest of some sort, with a blond girl winning it. After this, a dark-haired young woman (Laura Elena Harring) is being driven in a limousine when it stops and the driver takes out a silenced gun and tries to kill her, but a speeding car with celebrating youth rushes toward them; there is a horrific accident and the woman, although injured, is the only survivor. In shock, she makes her way down to Los Angeles and sneaks into an apartment. Later that morning, an aspiring actress named Betty Elms (Naomi Watts) arrives at the apartment, which is normally occupied by her Aunt Ruth and is shown into it by Coco, the manager. Betty is startled to find the woman, who has amnesia and calls herself "Rita" after seeing a poster for the film *Gilda* starring Rita Hayworth. To help the woman recollect her identity, Betty looks in Rita's purse, where she finds a large amount of money and a blue key. In another thread of the film director Adam Kesher (Justin Theroux) is casting the female lead in his next film and he is being commandeered by mobsters (the Castigliane brothers) to choose an unknown actress named Camilla Rhodes for the role. Kesher refuses, but his line of credit is withdrawn and he has to compromise.

The following interpretation will find detractors, but that the first half of the film is "real" is evidenced by Lynch treating Los Angeles and Hollywood with distinctly comic revulsion. Betty gets off the plane with an elderly couple, both from the city. She thanks them for making her comfortable and they beam incessantly, their ecstatic smiles not leaving their faces even inside their own car, ironically emblematic of the intense pleasure produced by their city! Los Angeles arrives ceremonially for Betty when she sees the banner welcoming her. Betty is from Deep River, Ontario, and Los Angeles electrifies her, but Lynch is laying it on extremely thick, substantiated by the savage view he provides of Hollywood and its milieu thereafter.

The Castigliane brothers are (like Mr Eddy in *Lost Highway*) uncouth; the studio official orders espresso for the older one, assuring him that it comes "highly recommended." What Luigi Castigliane does thereafter when he takes a sip of the espresso is to spit it out on a spotless napkin even as the official pleads with him about it being "the best quality." The meeting is inconclusive and it angers the mobsters but Kesher walks out, takes a golf club from his car, and smashes the Castiglianes' car before fleeing in his own vehicle. Kesher later relents through the good offices of a mysterious intermediary named "the Cowboy," who lives in a corral on a hilltop and advises him sagely, "A man's attitude goes some ways in how his life will be." Adam Kesher has also gone home in between to find his blond wife in bed with the pool cleaning person, who then beats him up. In retaliation, Kesher pours pink enamel paint on his wife's jewelry and the pool cleaner thinks, "That's no way to treat your wife, no matter what she's done."

These parts of the film—including the assorted doings of a bumbling hitman searching for a phonebook—are inherently comic and arguably satirical, implying a very real target, which is dealings in Hollywood. Even if this first part is the "imagining" of someone, we need to take cognizance of it being a nasty caricature of a physical Hollywood. Betty does an audition which goes very well and we get more satire of Hollywood scripts and casting agents and their doings.

Adam Kesher is in the same location casting the actress named Camilla Rhodes chosen by the mobsters with the chosen words, "This is the girl," and Betty catches his eye, suggesting that he might have chosen her if he had had the chance. Betty has however to hurry off because Rita has suddenly remembered a name "Diane Selwyn," who they soon locate in the directory and then seek out. Diane has moved to another apartment and the girl who is in her former apartment directs them there. We are shown the two *from the girl's viewpoint*, implying that what happened hitherto has been real. This other girl has had no other part in the story *except to witness this*. In Diane's apartment, however, they find a young woman's corpse, several days old.

There is now an interlude of sex between Rita and Betty and Rita wakes up in the middle of the night saying "Silencio" and insisting they drive to someplace in the far end of Los Angeles and they take a taxi to a club by that name. This sequence is electrifying—music is being played and the MC makes an announcement, "There is no band," and demonstrates that the man on the stage is only going through the motions of playing when they are hearing a recording. Lynch has the propensity to produce hypnotic effects but this segment is arguably symbolic: that all kinds of effects are produced even when there are no designated players, which I take to reflect upon the murky goings on in Hollywood that follow this segment and involve the girls.

Lynch has, earlier, put in an interlude involving two men in "Winkies" a café on Sunset Blvd, one of them telling the other about a frightening dream he has had, about an awful face behind the café and the men proceed there to actually see the "face." As in his other films, he puts in things that only add to the mood without implicating them in the plot, which, I propose, absolves us from regarding every episode as sharply interpretable. This "inexactness," as reiterated, implies that a close reading of Lynch's films may not be always rewarding. What matters is the overall design and not every frame or twist, as film students may search out.

We now move to Diane Selwyn's story and discover that Betty is actually Diane, a failed actress; Rita is actually Camilla Rhodes. At Camilla's invitation, Diane attends a party at Adam's house on Mulholland Dr At dinner, Diane (previously Betty) indicates that she came to Hollywood from Canada when her Aunt Ruth died and left her money and she met Camilla at an audition for "The Sylvia North Story." Adam's party has many of the characters encountered hitherto including the earlier Camilla Rhodes and the Cowboy; Coco, who showed Betty into her aunt's apartment, is Adam's mother. Adam announces his engagement to Camilla, sending Diane into paroxysms of jealousy. Later, Diane pays the hired killer from the earlier section to eliminate Camilla Rhodes but also kills herself, which ties in with Diane's rotting corpse being discovered by the girls. The film ends with an unknown woman on the screen saying, "Silencio."

The first part of the film that I described as "real" has several mutually exclusive threads, working against the possibility of it being the self-justifying fantasy of a single person (Dianne who imagines herself as Betty), and its reconstruction is also too trite an exercise for a major filmmaker to set his audience. My own sense of *Mulholland Drive* is that Lynch is dealing with the compulsions in Hollywood that make people act against their own judgments, moral or otherwise, because of an impersonal predatory system long institutionalized. Betty helping Rita but also (as Diane) and ordering her liquidation (as Camilla) is pointing to the erasure of individuality, a cherished component of Hollywood's mythology.

If all people placed in the same situations act in identically shameful ways, where does individual moral choice come in? Once this is admitted, who Betty or Camilla really is becomes redundant. The fluidity of identity implied in the party scene at Adam's makes a mockery of individuality, also lampooned in a different way in *Wild at Heart*. If Lynch has been parodying Hollywood genres, *Mulholland Drive* is arguably the culmination where Hollywood itself is satirized by problematizing one of its most cherished mythologies, that of the stable individuality. *Lost Highway*, we may note, deals with an affiliated notion as well, which was moral continuity in the individual.

The episode in "Silencio" where there are effects without players points to the absence of individual "authors" in Hollywood's goings-on, an institutionalized mechanical process—like a musical recording—but unholy and without a remedy. The term "silencio" only means "silence" which could refer to the way that none of these features of Hollywood are ever noted or acknowledged. The critique of Hollywood in the film (unlike in *The Player*) deals with bizarre aspects like the gangland pressures through the Castiglianes and the involvement of "the Cowboy" as mediator/influencer that appears novel but *The Godfather* (1973) did hint at it in another way in the episode involving the Corleones, Johnny Fontane and Woltz. The uncouth Castiglianes are arguably a corrective applied to the flattering picture of mobsters in *The Godfather*.

Lost Highway and *Mulholland Drive* deal with the issue of identity and both are nominally murder mysteries.

> What does the moviegoer expect to find out when asking "whodunit?" Simply speaking, when narrative cinema poses the question, the answer is likely to be the murderer's name. Looking for a murderer's name already involves philosophical premises concerning both the nature of the subject and the nature of the world. With regard

> to the subject, a name is assumed to stand for one's identity. That is, it posits the existence of a stable, continuous entity within the solid physical boundaries of the body; an entity acting and existing within the limits of the phenomenal, subjected to natural laws of time and space.[23]

Habitually, Lynch's interpreters turned to psychology, the studies examining the unconscious aspects of human experience, and read the Lynchian world as unleashing the wild, unruly side of the human psyche. Therefore, most of the scenes are regarded as fantasies, which do not occur in the cinematic real life or real world but in the mind of one protagonist or another. Since mental images are not ontological categories, they are not acknowledged as real events.[24] The above writer, instead, deals with the fluidity of identity and goes on to relate this fluidity of the identity to "Advaita Vedanta," an idealistic school in Indian philosophy. But my assessment is that the film is not strictly *propagating* that philosophical viewpoint but interrogating aspects of the dominant mythologies created by Hollywood that privilege individual choice and identity.

The ferocity with which Lynch lampoons Hollywood and the habitual goings-on in it in *Mulholland Drive* provides evidence of the film being satire more than a philosophical reflection. Lynch's antipathy toward Hollywood is an emotion exhibited *toward something real* that impinges upon his own life and not primarily a philosophical reflection. But if he is constructing an elaborate plot as a "thought experiment" (as in *Lost Highway*) to interrogate a dominant mythology, he could still be influenced in this undertaking by Advaita and his occult imaginings may have antecedents. The personal fantasy/nightmare was something he used in his first film *Eraserhead* but it subverted comforting sentiments around childbearing and the nuclear family.

Twin Peaks (1990–2017)

Twin Peaks, a television series created by producer Mark Frost and David Lynch, premiered on April 8, 1990, on ABC, running until its cancellation after its second season in 1991. Before Lynch and Frost finalized their *Twin Peaks* concept, they watched the 1957 film *Peyton Place*, which explored the hidden secrets of a New England small town. There are a number of details in the film that remind one of *Twin Peaks*. *Peyton Place*'s lead male character is a thoughtful outsider figure who is introduced when driving his car toward the small town where he takes on a new job. *Peyton Place*'s major employer is a mill. A man forces himself sexually on his blond high-school-age stepdaughter. Much of the story centers on teenagers going to high school.[25]

In Season 1 of *Twin Peaks*, local logger Pete Martell discovers a naked corpse wrapped in plastic on the bank of a river outside the town of Twin Peaks. Washington

Sheriff Harry S Truman, his deputies, and doctor Will Hayward identify the body as high school senior and homecoming queen Laura Palmer. A badly injured second girl, Ronette Pulaski, is discovered amnesiac. FBI Special Agent Dale Cooper is called in to investigate. Cooper's initial examination of Laura's body reveals a tiny typed letter "R" inserted under her fingernail. The series combines local conspiracies and dreams freely, and dreams contribute that characteristic occult element familiar from Lynch's work. Dreams then become a legitimate way of disentangling the mystery, even by the FBI, which could be seen as a parodic turn.

Season 2's later episodes were passed on to other directors and began to see its ratings dropping; Lynch was persuaded to reveal who Laura Palmer's killer was, a decision he later regretted, but Season 2 was nonetheless stopped with Lynch directing the final episode, trying to turn it into a cliff-hanger. Following the success of *Wild at Heart*, Lynch returned to the world of *Twin Peaks*, this time without Mark Frost, to create a film that acted primarily as a prequel but also, in part, as a sequel. The result, *Twin Peaks: Fire Walk with Me* (1992) revolved around the last few days in the life of Laura Palmer, and was "darker" in tone than the television series, and dealing with such topics as incest and murder—but going so overboard in its fantastic and occult elements as to be impossible to relate coherently. The film was a commercial and critical failure in the United States although praised by some critics.

Twin Peaks: The Return (2017)

Twin Peaks re-emerged in Season 3 in 2017, this time entirely directed by Lynch himself and concluded in a way that would prevent further continuations. This series, in my view, is a significant cinematic achievement and what follows is an attempt to make some sense of it. Given Lynch's propensities and my own approach declining to classify the "real" and "imagined," "making sense" of it cannot evidently be in terms of reconstructing the plot, and I will restrict myself to speculating upon the nature of the achievement, in the light of his earlier work.

As already explained, the 1990s *Twin Peaks* series (S1 and S2) was about the unexplained murder in the town of Twin Peaks in Washington State, of Laura Palmer, the homecoming queen, whose naked corpse is found, wrapped in plastic. FBI Special Agent Dale Cooper (Kyle MacLachlan) is called in to investigate and it is discovered that Laura Palmer was into drugs and prostitution. Where this beginning should have taken the story into the expected route of clues, motives, and connections, it goes instead into occult (often incoherent) territory with Dale Cooper having strange dreams and visions that he uses to try to solve the mystery around Laura Palmer's death.

The first two episodes of *Twin Peaks* Season 3 were part of a preview and they set the tone for what follows. To describe S3, it unfolds in a series of events taking place around Twin Peaks, in Buckhorn (South Dakota), New York, and Las Vegas. There are realistic bits dealing with people conducting themselves as people do, though the way they are enacted gives them a comical edge that suggests social satire. The most striking happening in the first episode is perhaps the one set in New York in which a young man Sam is watching a glass box intently over a long period of time, capturing what is happening on film and storing the slides away carefully. The glass box is connected to a highly complicated electrical apparatus and the place is fiercely guarded. No one is normally allowed inside but Sam's girlfriend Tracy comes in with coffee when the security guard is absent. Sam knows little about his "work" except that this is an experiment funded by a billionaire; he has seen nothing so far but the man he had replaced saw "something" happening in the glass box although Sam does not know what.

The segment just described is riveting but comic at the same and the reason is that its effect is parody. We are aware of SciFi films in which people watch for some phenomenon—usually celestial as in *Contact* (1997)—and Lynch is arguably parodying it. To make it funnier, the opening in the glass box is from where Agent Dale Cooper (apparently trapped in another dimension) enters the world! This is Lynch again subverting a public mythology through a personal fantasy. In another segment pertaining to Buckhorn a woman in an apartment block smells something foul from the neighboring apartment of librarian Ruth Davenport and calls the police. When the police enter the apartment, they find Davenport's severed head aligned on her bed with a nameless male body. When Davenport's apartment is searched for prints those found everywhere belong to the local school Principal Bill Hastings, a classmate of the police officer in charge. This again feels like a parody but of another genre—the police procedural.

It is difficult to explain the effects these segments have on us because they can be so ludicrous while seeming to be in dead earnestness. My sense is that the solemnity with which they are enacted conceals the fact that they are essentially nonsensical—as most parody is making nonsense out of something deeply serious. In other parts of Season 3, Lynch brings in issues like superheroes, nuclear explosions, the occult beliefs of Native Indians and secret military experiments during the Cold War—associated with the top-secret "Blue Rose"—not to mention the inner workings of the FBI, which serves as an undercurrent. The deaf Gordon Cole played by Lynch is a continual source of amusement as is his tendency to jump to occult explanations before even comprehending what is happening. Still, Lynch is not simply lampooning generic films of various kinds in these segments but offering something else, evidently deeper.

All film genres are based on mythologies in the public space and society functions with assistance from them. The mythologies help to fill gaps in public understanding and hence lend stability to any social structure. Motifs ranging from the civilizing influence of the early pioneers (embodied in the western) to monogamous heterosexuality (found in every genre, but especially in romances) are illustrations of such mythologies. The mythology around the alien landing is specifically associated with a global superpower that does not see equals and therefore looks outside—whether for company or stronger adversaries. Most of Lynch's films, as I have argued, have worked with these mythologies, interrogating aspects of them but also going off to raise affiliated philosophical questions like those involving individual identity.

My proposition here is that the genre films, by "mythologizing" social experience, have increasingly obscured the fact that there is not enough understanding of a subject; they make it seem that opaque matters like communications technology or secret military doings are within the grasp of the public and remediable—and they respond with mythical "feel good" solutions to actual problems that defeat solutions, and this last aspect has become notable in the recent past since it is now that such matters are becoming opaque. *Iron Man*, for instance, responds to the entanglement in Afghanistan by concocting an invention with the potential to help America overcome military adversities. The swamp creature turned into a "military asset" in *The Shape of Water* (2017) could be a prisoner from Guantanamo Bay from whom something is still to be extracted. Afghanistan and Guantanamo Bay involve political decisions that the public is helpless about despite the United States being a democracy, and cinema hence steps in.

What Lynch appears to be doing in *Twin Peaks* Season 3 is to subvert each genre he toys with, by leaving things deliberately mysterious—like how Bill Hastings' fingerprints got all over Ruth Davenport's apartment or what the experiment with the glass box and watched over by Sam is—and then offering outrageous "occult" explanations that satirize the public's claim to full knowledge of its milieu and its complacent sense that all problems have publicly admitted remedies. A highly advanced technological society like the United States depends on the public's obliviousness to what is going on since widely held knowledge might become problematic. It could be argued that the lack of public knowledge has a corresponding "silence" of some sort; i.e. the "Silencio" alluded to in *Mulholland Drive*. Lynch, by mocking the access to such knowledge and their remedies, is providing a critique of his own society's smug beliefs; he is shooing them away with his own private ghosts as he did in *Blue Velvet*. But with so many generic conventions being lampooned, the underlying satirical tinge is more palpable.

Lynch's films being a medley of different genres (like Tarantino's) has led to the epithet "postmodern" being used in connection with his work, but there is a difference that rests in Lynch's work (unlike Tarantino's) being interpretable as satirical or parodic. *Twin Peaks* Season 3 parodies genres rather than pastiches them. Pastiche is "blank parody" or "parody that has lost its humor"[26] and people do not usually laugh when Tarantino pastiches "time management" education in a segment from *Pulp Fiction* ("The Bonnie situation"). *Twin Peaks* Season 3 can be funny since he parodies rather than pastiches genres. Lynch's parody of genres ultimately points to social satire since it subverts the mythologies on the basis of which society subsists. Satire is normally obvious but Lynch's approach—with so many occult elements—makes it elusive and it remains ironical and rich in hermeneutic possibilities rather than openly satirical. The treatment of identity in *Lost Highway* and *Mulholland Drive* can even lead to philosophical speculation.

A last aspect of *Twin Peaks* Season 3 that needs commenting upon is Lynch's use of popular music as performance. The second episode in the series concludes with "Chromatics" performing "Shadow" in a Twin Peaks bar. The performances thus rendered in the series are all conducted in the same "Bang Bang Bar" at the conclusion of episodes, with the characters watching, and one notices how restful they are. The lyrics are banal but the banality quietens the disturbances that the film has made us undergo. One recollects Dorothy Vallens singing "Blue Velvet" in Lynch's film and the unspeakable Frank Booth listening with tears streaming down his face, Henry pacified by the girl in the radiator singing "In heaven, everything is fine" and the people in the "Bang Bang Bar" could be responding with similar emotions there.

Twin Peaks Season 3 is not meant for passive viewing like much of television; we have to "cope" with it. If "Shadow" helps us cope it is in the company of the characters who see no end to the incomprehensible happenings in their lives. This leads us to another question—whether popular musical performances in an advanced technological society like the United States are, likewise, not pacifying the public, helping them cope with matters also complex and incomprehensible, perhaps more so than David Lynch can make them. Aki Kaurismaki, we recollect, also proposed something similar when he treated popular music as performance as a palliative for working-class distress.

Since Lynch and Altman have both worked with genre cinema, subverting or interrogating the dominant mythologies manifested in them, it would be helpful to distinguish between their respective approaches. Altman, I propose, is "realistic" in the sense that he is trying to lay bare or expose the mythology; both his generic subversions like *McCabe and Mrs Miller* and his arthouse films like *Nashville* do this although in different ways. David Lynch, on the other hand, advances his own elaborate—and usually fantastic—fictions as antitheses to dominant mythologies, and the bizarre excesses in his films are subversions of the genres tied to comforting mythologies—like Sailor smashing an adversary's skull as an expression of his "individuality and personal freedom." The fact that Lynch's fantastic visions can be interpreted separately—philosophically or psychoanalytically—cannot obscure the fact that they are aligned against the dominant mythologies, most often those perpetuated by Hollywood. Mythologies are, broadly, widely believed public narratives and Lynch's occult fictions are often like "thought experiments" falsifying them, his "falsifications" apparently a debunking, although these occult fictions may also be interpreted for independent philosophical significances.

NOTES

1. K. George Godwin, Review: *Eraserhead* by David Lynch, *Film Quarterly*, 39.1 (1985), 37.
2. Louis Althusser (*Lenin and Philosophy and Other Essays*) described institutions like family, school, and church as "ideological state apparatuses" (as against the police and the military which he called "repressive state apparatuses") since they exert coercive pressure upon the individual to conform socially/politically to state authority. Louis Althusser, "Ideology and Ideological State Apparatuses," *Lenin and Philosophy and Other Essays*, trans. by Ben Brewster (New York: Monthly Review, 1971), pp. 142–44.
3. "Pro-life" platforms should perhaps be recoiling in horror at the way the baby is represented.
4. This has been sensed by critics when they see it as a comedy without humour. "At its most accessible level, the film seems to be a strange domestic comedy in which the little annoyances of daily life are blown up to monstrous proportions, with poor bemused Henry stumbling through it all, trying his damnedest to appear inconspicuously normal in situations where he is not certain just what 'normal' is. But this humor is not funny, because it is wedded to dark, bleak imagery." K. George Godwin, "Review: *Eraserhead* by David Lynch," *Film Quarterly*, 39.1 (1985), 38.
5. Slavoj Zizek, *The Art of the Ridiculous Sublime: On David Lynch's Lost Highway* (Seattle: University of Washington Press), 2000.
6. Todd McGowan, "Sacrificing One's Head for an Eraser," in *The Impossible David Lynch* (New York: Columbia University Press, 2007), p. 26.
7. McGowan, p. 27.
8. Todd McGowan, "The Integration of the Impossible Object in *The Elephant Man*," in *The Impossible David Lynch*, p. 49.
9. Merrick being virtually spat upon (it happens when he is onstage), "bearing his cross" without complaining and his dead mother's beauty (Merrick calls it "angelic") do suggest a religious thread in *The Elephant Man*.
10. Nadja Durbach describes the work as "much more mawkish and moralising than one would expect from the leading postmodern surrealist filmmaker" and "unashamedly sentimental." Nadja Durbach (ed.), "Monstrosity, Masculinity, and Medicine: Reexamining 'the Elephant Man'," in *The Spectacle of Deformity: Freak Shows and Modern British Culture* (Berkeley: University of California Press, 2009), p. 35.
11. David Lynch, *Lynch on Lynch*, ed. by Chris Rodley, revised edn (New York: Faber and Faber, 2005), pp. 11–15.

12. Todd McGowan, "Fantasizing the Father in *Blue Velvet*," in *The Impossible David Lynch* (New York: Columbia University Press, 2007), p. 91.
13. "That is to say, when we fantasize about our ideal, we fantasize simultaneously about the threats that imperil that ideal." McGowan, p. 93. I contest the notion that the small-town ideal is "fantasy" since that would tend to make it private. It is more in the nature of myth rather than fantasy.
14. *Blue Velvet* is often termed "neo-noir" and that would place Frank Booth in the same category as Noah Cross from Polanski's *Chinatown*. The plot issues resolved in neo-noir are left entirely unresolved in *Blue Velvet*. Both villains may be "evil" but there is less logic to Frank Booth, his motivation is kept opaque, and his extravagance is more in the line of the Joker in *Batman*.
15. Todd McGowan, "Fantasizing the Father in *Blue Velvet*," in *The Impossible David Lynch*, pp. 90–91.
16. Some important artists here would be Hieronymus Bosch, Francisco Goya, Pieter Brueghel, and Francis Bacon.
17. See Endnote 13, Chapter 4.
18. A way in which this might be approached is through Mikhail Bakhtin's notion of "dialogism," the rejection of an authoritative narrative voice intended to convey a single thematic message. Lynch is arguably disrupting the unity of the film by introducing pluralistic elements not in harmony/consonance with each other but working at cross-purposes, thereby undermining the unified sense of nationhood as well. Michael Dunne, "*Wild at Heart* Three Ways: Lynch, Gifford, Bakhtin," *Literature/Film Quarterly*, 23.1 (1995), 6–13.
19. David Lynch, *Lynch on Lynch*, ed. by Chris Rodley, p. xi.
20. Making sense of a narrative is a process following its reading. It corresponds to "narrativity" or the active construction of the story by the reader/spectator from the fictional/visual/aural data made available. Robert Scholes, "Narration and Narrativity in Film," in *Film Theory and Criticism: Introductory Readings*, ed. by Gerald Mast and Marshall Cohen, 3rd edn (New York: Oxford University Press, 1985), pp. 392–93.
21. Here is Slavoj Zikek on the subject: "Patricia Arquette was therefore right when, in an effort to clarify the logic of the two roles she was playing, she produced the following frame of what goes on in the film: a man murders his wife because he thinks she's being unfaithful. He can't deal with the consequences of his actions and has a kind of breakdown in which he tries to imagine an alternative, better life for himself, i.e. he imagines himself as a younger virile guy, meeting a woman who wants him all the time instead of shutting him out, but even this imaginary life goes wrong—the mistrust and madness in him are so deep that even his fantasy fails apart and ends in a nightmare." Slavoz Zikek, *The Art of the Ridiculous Sublime: On David Lynch's Lost Highway* (Seattle: University of Washington Press, 2000), p. 20.
22. Maxwell's Demon is a thought experiment that appears to disprove the second law of thermodynamics. It was proposed by the physicist James Clerk Maxwell in 1867. https://www.britannica.com/science/Maxwells-demon. Accessed August 22, 2024.

23. Ronie Parciack, "The World as Illusion: Rediscovering Mulholland Dr and Lost Highway through Indian Philosophy," in *The Philosophy of David Lynch*, ed. by William J Devlin and Shai Biderman (Lexington: The University Press of Kentucky, 2011), p. 78.
24. Parciack, p. 80.
25. The relationship with *Peyton Place* is noted by Greg Olson. Greg Olson, *David Lynch: Beautiful Dark* (Lanham: The Scarecrow Press, 2008), pp. 266–67.
26. Frederic Jameson, "Postmodernism and Consumer Society," in *Postmodernism and Its Discontents*, ed. by E. Ann Kaplan (London: Verso, 1988), p. 16.

Afterword: Targets and Strategies

When I began planning this book, the primary impetus was to write about a certain kind of sensibility in cinema that I felt attracted to and thereafter to identify the filmmakers who worked best in that mode, with films that could often be termed complex and ambiguous. Irony is often a tone adopted by such a sensibility and the filmmakers using it are hardly few in number. "Distance" is perhaps as common a phenomenon in artistic expression as empathy but I was seeking to identify major artists in cinema who had a vision of the world that was manifested in the strategy of irony, and used that over a whole career comprising a large number of films. There are a number of filmmakers who are sardonic when dealing with their subjects, and some recent ones who I might have considered—to bring the book up to date—might have been the Romanian director Radu Jude (*Aferim!*, 2015), the Austrian Ulrich Seidl *(Paradise: Love*, 2012), and the Swedish filmmaker Ruben Östlund (*Triangle of Sadness*, 2022). All of them are interrogating their own societies and their pasts, with irony/satire as the chief strategy.

But the older filmmakers I finally stuck to are more exploratory and complex in their expression. If I were to identify the aspect of their films where this is manifested, each of these selected filmmakers constructs intricate narratives with characters and situations while the newer ones just named rely on a tone, sometimes counterposing a transparent situation with the ironically unsuitable rhetoric—as for instance the rhetoric of love used by Seidl's Austrian sex tourists and the locals in Kenya. It could be argued that the interpretable elements in a narrative are related to the number of socio-political categories represented; the complexities in the storytelling would naturally lead to more meaning being made out of it. To consider Buñuel's *El*, it illuminates different facets of the bourgeoisie, the church, and patriarchy across classes by exploring relationships between categories of people. Balabanov's *Cargo 200*, in dealing with the collapse of the USSR, represents the state through the militia, the military fighting in Afghanistan, the *nomenklatura*, the emerging business class of the post-Soviet era alongside a philosophically cogent account of the materialist position as articulated in the Soviet universities. Seidl's film, while being deeply ironical, deals with only two social categories and has therefore a less complex take on the situation it is depicting, primarily implying the insincerity of the rhetoric. An intricate narrative perforce introduces a larger number of signifiers to draw the spectator into the interpretive act—necessary for the irony to be detected. Satire as a genre is more obvious and does not rely much on interpretation.

The selection of targets

Based on my arguments, it would be easiest to begin with the targets selected by the respective filmmakers since what marks them out is the mockery or irony they employ. These targets can be categorized as under: (a) the public, especially those of a particular society, for characteristics they are seen to possess, (b) a political system for the kind of relationships it engenders, (c) a social system, its authors and conventions, (d) the dominant mythologies prevalent in the socio-cultural system, and (e) the relationships between different categories within a given social structure. Needless to add, there could be more than one target in a filmmaker's selected work.

Humankind or national societies

Kubrick is the only one among the filmmakers to pick humankind for his irony or his barbs and largely for its capacity for violence; but associated with this is the notion of human-devised systems failing. His anti-war films should be singled out here when he targets the upper echelons in the military (*Paths of Glory*) for sending soldiers under them to certain death but with an elaborate strategy underlying it. Humankind puts its destructive urges unwittingly into a supercomputer in *2001: A Space Odyssey*, while its members acquire the characteristics of automatons. Regardless of how the film concludes, HAL's malfunctioning in the film is symptomatic of a failed human system. *Dr Srangelove* is, similarly, not an attack on the military establishment but on humankind's destructive urges that find manifestation in the defense systems on both sides.

The same observation about failed systems can be detected in *Full Metal Jacket* where the intention is to create "perfect killing machines." But in all of his last three films, the target is more especially the United States, its military establishment in this film and its postmodernity in *The Shining*, where *The Overlook Hotel* is a failed system. In *Eyes Wide Shut*, he proposes that America itself is a failed system managed by a mysterious plutocracy riding on the all-pervasive myth of equality among its citizens. Tom Cruise is used in the film to represent "Mr Everyman" and he is endowed with the moral squeamishness that Verhoeven also targets America for, in his films made there.

Verhoeven's *Turkish Delight* can be termed an attack on hedonism in Dutch society, also once targeted by artists like Hieronymus Bosch, but in his American films he targets specifically American socio-political targets although filtered through media representations, which implies that it is media representations in America—advertising, reporting, and the glamor industry, chiefly—that are his primary targets. Still, he does not see the media as separate from the country and its society.

Fassbinder is another filmmaker who targets his own society and especially its people for their political errors—usually not catching the larger picture and the meaning of their political choices. I would trace this distaste to the enthusiasm with which Germans welcomed Nazism and Hitler, the consequences of which resulted in a trauma that has still not left Germany. In his TV film *Jailbait* (1972), not included here, a wife criticizes her husband for being a Nazi sympathizer since "the Nazis also made mistakes!" In *The Marriage of Maria Braun*, the woman protagonist dying in a gas explosion while being personally blameless in the period of Germany's resurgence is a hint of the "original sin" that all Germans must continue to pay for.

Political systems

The biggest critic of a political system among the filmmakers examined is Aleksei Balabanov whose films invoke Communism's failed utopia, attributed chiefly to human irrationality. I noted how that is represented by an absence in *Happy Days* where the protagonist has no memory of the entire period. The film is set in St Petersburg which became Leningrad under the USSR and reverted back to its original name in 1991 and this return to original nomenclature reflects a period best forgotten. Balabanov also uses silent cinema—one of the most enduring contributions of the Soviet Union to culture—in his mockery when one of his protagonists becomes a famous filmmaker making pornographic films in the silent era (*Of Freaks and Men*). The trope of silent cinema recurs in *Morphine* but, in his films set in the USSR and after, he uses genres (the gangster film, the horror film, the revenge film) to also talk about the collapse of an unviable utopia that led later to large-scale criminality. The Yeltsin period and the rise of business/criminality are a key motif in all the films except *Happy Days* and *Morphine*. The deliberate polyphony in his films essentially manifests the failed system.

Aki Kaurismäki is another filmmaker deeply critical of the "capitalist" political system of his own country, and there is an indication that he is a socialist at heart and a former Soviet sympathizer. I partly interpret his "dead-pan-emotion" filmmaking style as politically indicative of ordinary citizens being at sea in their societies—so perplexed are they that their faces register nothing. This is a way of distinguishing content wise between Robert Bresson and Kaurismäki who are both "minimalist." Since Bresson's focus is directed "inward," his inexpressive countenances indicate people not fully in touch with themselves and their motivations, while in Kaurismäki, who is political, it is indicative of people unable to comprehend the (nominally democratic) system that envelops them. Kaurismäki's later films depart from this and the admission of dogs into narratives (since dogs register emotions) is arguably indicative of his reconciliation with the political system—after the fall of the USSR, in which he apparently had covert faith.

The social system and its authors

Among the filmmakers selected the only one who worked in several political regimes—Spanish, French, and Mexican—was Luis Buñuel but the target of his irony and satire remained the same each time—the bourgeoisie and the Church. Buñuel was basically targeting structures and social conventions created by the upper classes with the assistance of the Church; he saw them to be deeply one-sided, not fulfilling the supposed moral purpose they claimed. As an instance, Joseph in *The Diary of a Chambermaid* rapes and murders a little girl but protests the charge by giving his devoutness and patriotism as evidence. Buñuel is implying that neither one's devotion to God nor loyalty to one's country can be termed a moral quality. As part of his antagonism to organized religion (Catholicism), Buñuel denied God in a large number of his films by subverting the teleology inherent in the narrative; he declined to play "God." *The Phantom of Liberty*, while partly retaining the same targets, endeavors to examine the way social conventions embedded in the structure make us act in everyday life rather than moral or social purposes—to which prescribed conduct is traditionally traced. Patriarchy, another target, is seen as embedded in bourgeois social conventions.

Mythologies

The two filmmakers who are both primarily concerned with subverting mythologies and largely through their interrogation of Hollywood genres are Robert Altman and David Lynch. A large part of Altman's output is given to generic revisions in which certain cherished notions with regard to the social past are interrogated—not to make a counterclaim but to question its construction. As an instance is his notion of when America "happened"; where the western normally sees as happening when civilization was brought to a savage land (as in John Ford's films), Altman sees it as happening (*McCabe and Mrs Miller*) when the monopoly house destroyed the small business. In *Nashville*, he is doing the same with another kind of national mythmaking—the world of Country Music. *3 Women* deals with people constructed by consumer advertising and his Millie's life could well become an instantiation of Roland Barthes's *Mythologies,* in which many of the essays explore the myths perpetrated by the magazine *Elle*.

Lynch is also deeply preoccupied with Hollywood and its genres and the split in his films is not between "reality" and his private fantasies but between Hollywood's reassuring generic representations and the dark personal fantasies that he combatively constructs. *Eraserhead* is the only selected film in which Hollywood does not apparently mediate but notions important to its fictions—like "life," God, and parenthood—are savagely interrogated. I am even tempted to see the fantastic elements as deliberately employed to conceal taboo notions like infanticide, which might be impossible to deal with realistically. The "Lynchian fantastic" in *Blue Velvet* is not entirely horrific as it is taken to be but is even comic because of its over-the-top representation of "evil" as cliché (sexual perversions and the whispered word "drugs") instead of seeing evil as simply moral choice and wrongdoing. Lynch also uses the fantastic to interrogate the notion of individual identity, which is a key notion in Hollywood films. His television series *Twin Peaks: The Return* (S3) actually works sequentially with a number of genres and he uses the fantastic ("the occult") as a way of interrogating the "unacknowledged" (like the ultimate purposes of technological advances, the rationale for military decisions) in a democratic society where the citizen should supposedly have access to every kind of information.

The relationships between different social categories

Although mockery of a system, a society or a mythology would imply exploring the relationships between social categories, that is not what a majority of the filmmakers attempt. Buñuel is evidently exploring relationships—like between male employer and domestic servant—that throw up patriarchy and other asymmetric relationships usually dictated by the political structure. For example, the position of the woman in *El* owes ultimately to social structures in which the Church plays a mediating part (Louis Althusser's "Ideology and Ideological State Apparatuses"). But one does not find the exploration of social relationships particularly prevalent in Kubrick's films or Altman's or Lynch's, none of who also pays much attention to gender issues.

Among the filmmakers exploring political systems, Balabanov represents social categories and the relationships between them through transactions at the individual plane. The contaminated morphine in *Morphine* on account of the protagonist replacing it with something else kills their neighbors but, they being prosperous kulaks, it will not get him into trouble. *Cargo 200* deals with the relationship between state power (Capt. Zhurov) and the *nomenklatura* (Angelika, the local party chief's daughter). The regional/racial identity of the protagonist (Yakut) within the former Soviet army plays a large part in what happens to him in *A Stoker*. But this relationship between distinct social categories is less in evidence in Kaurismäki's films (excepting *The Match Factory Girl*) and he usually deals with the working class's position in the political/social system and not its relationship with the other classes.

Fassbinder is dealing with different classes in his films and their relationships obviously interest him—like the white working woman and Gastarbeiter in *Fear Eats the Soul*. But it seems secondary or subordinated to his primary preoccupation which is to bracket all Germans together on account of their shared past. I would argue that in *Fox and His Friends* also, although it is about class relationships, the emphasis is on the protagonist's naivete especially when Fassbinder presents his pathos as cliché—his corpse being robbed of its last coins by two schoolchildren. Verhoeven, the remaining filmmaker, has no interest in social relationships and they feature only when they are implicated in the genres he is working with—like those in *RoboCop*. Kubrick, Altman, and Lynch also do not show themselves as interested in exploring social relationships.

Strategies to indicate irony

The other aspect that needs exploration apart from the choice of targets is the common strategies employed by the filmmakers to suggest irony or satirical purpose since that too could have wider connotations outside the work of the selected filmmakers. Regardless of whether most of his/her audiences might miss it, irony must still allow itself to be registered and that evidently calls for a strategy.

Understatement or cliché

Understatement and cliché represent the polarities in the suggestion of the ironic but both imply a gap between expectations in the portrayal of a situation and its actual representation. That they fulfill the same purpose in implying irony is suggested by Fassbinder who used minimalist understatement in *Katzelmacher* and then went on to use melodrama as cliché in films like *Fox and His Friends* and *Fear Eats the Soul*, although even the latter films are more understated than the melodramas like *All that Heaven Allows* that inspired them. Fassbinder also used an overblown film style spelling artifice especially in his use of lighting and color in *The Marriage of Maria Braun* and *Lola*, melodramas set in the period of Germany's economic resurgence.

Two other filmmakers who use understatement are Luis Buñuel and Aki Kaurismäki although differently. Buñuel does it by using a realistic filmmaking style while creating situations that can even be termed "impossible" as for instance the conduct of the parents when their child is "missing" in *The Phantom of*

Liberty. That is why we are constantly surprised by what happens in his films: it is as though people contrive to take the most impossible things in their stride. Other instances are the conduct of the man toward the "two women" in *That Obscure Object of Desire*, the casual way in which Joseph's rape and murder of the little girl in *Diary of a Chambermaid* does not cause shock even in Celestine but only a desire to inflict punishment anyhow. *The Exterminating Angel* would be another instance though I only touched upon it briefly.

Kaurismäki uses a deadpan style and his people do not emote conspicuously in the horrendous situations they are placed in, and they carry on as though with fortitude toward adversity. The frozen Leningrad Cowboy finally thawing, climbing out of his coffin and performing music with the troupe and being accepted is an extreme instance. If in Buñuel the understatement seems indicative of acceptance in the negative sense—people capable of any perfidy without batting an eyelid—it is largely positive in Kaurismäki's films, they being able to suffer unemotionally. Even Liris poisoning everyone impassively in *The Match Factory Girl* is because of her (justified) rage at the world and her impassivity is contrasted with what she endures.

David Lynch's fantastic aspects often verge on cliché ("drugs" in *Blue Velvet* as emblematic of evil) but not always, and it is also the lampooned mythology (the warmth of Lumberton, Sailor's valorization of his personal freedom in *Wild at Heart*) that is treated as cliché. The portrayal of the FBI with their black suits in *Twin Peaks* Season 3 and the quick references to genres like the police procedural are also heavily dependent on our recognition of cliché. But Lynch's over-the-top fantasies often overdo the cliché as for instance the sick baby in *Eraserhead* as well as its drab industrial milieu. My own proposition is that many of the "fantastic" elements in his work are not recognized as clichéd reality because they are so extravagant, like the mobsters, the assassin, and the Cowboy in *Mulholland Drive* that could still have real counterparts in Hollywood—financiers, enforcers, and influencers—implicated in film production.

"Expressionist spaces"

Many of the filmmakers often use spaces in an Expressionist way as a strategy aimed at pursuing their ironic purpose. The Overlook Hotel as representing American postmodernity in *The Shining*, the small town where Millie and Pinky like in Altman's *3 Women*, the urban milieu in *Eraserhead*, emptied-out St Petersburg in *Happy Days* and *Of Freaks and Men* would be instances, as would the empty industrial landscape in Balabanov's *Cargo 200*. The protagonists are placed in that milieu and help contextualize the discourse—the socio-political situation referred to. Henry's family life being in a more ordinary space in *Eraserhead* would have eliminated the satirical element and we could not have accepted the horrific conclusion. Verhoeven is another filmmaker pertinent here but he goes about creating entire worlds because of his futuristic aims. Still these futuristic worlds are not plausible (as in *2001: A Space Odyssey*) but expressive. *Basic Instinct* plays up San Francisco's glamor to make it suitable for the overblown erotic content. Even his Las Vegas in *Showgirls* is a distorted, showbiz version of the actual city, Las Vegas turned into a "hyperreality" as Jean Baudrillard might have had it.

"Disjunctive" music

A feature in common between many of the filmmakers is their use of music that might be normally considered inappropriate for the situation depicted—Kubrick's use of classical music (Penderecki) in a horror film, *The Blue Danube* in *2001* and Rossini in a rape scene in *A Clockwork Orange*, Balabanov's preference for cheap pop used often in tense sequences as in *Cargo 200* when Zhurov abducts Angelika, Kaurismäki's constant use of raucous diegetic music played in the saddest cafes and bars, and David Lynch's use of diegetic pop music seemingly as a means of emotional pacification. The last is a little different from the others and Frank Booth listening teary-eyed to Dorothy Vallens singing "Blue Velvet" has us wondering. But overall, the music may not be simply in opposition to the mood of the sequence although Kubrick often suggests that. But one could propose that the fact that the music often imposes a mood at odds with what is happening/has happened/will happen casts some doubt on the filmmaker's intent. It is partly this strategy that helps us to read parts of each film as irony.

Irony and satire

This book is basically about irony and the satirical impulse but it is preoccupied with the interpretive act and is hence primarily dealing with irony. As already indicated, satire does not often need interpretation and the intended satires in cinema named in this book are perhaps only a few—Kubrick's *Dr Strangelove*, Fassbinder's *Katzelmacher*, Altman's *M*A*S*H*, and (perhaps) Verhoeven's *Starship Troopers* although the last film can also be enjoyed unabashedly as an action romance. Where that film chiefly offers itself for interpretation is in its target since it is broadly taken to be "anti-militarist" without much recognition of what "militarism" means. This book deals with only eight major filmmakers; but their targets and the strategies employed—to deliberately say something while implying something else—could become valuable in relation to all cinema—since (as already noted) irony and satire are not popular today, when there appears to be a resolute tendency toward the promotion of empathy in cinema.

Bibliography

Althusser, Louis, *Lenin and Philosophy and Other Essays*, trans. by Ben Brewster (New York: Monthly Review, 1971).

Aumont, Jacques, Bergala, Alain, Marie Michel, and Vernet, Marc, *Aesthetics of Film*, trans. by Richard Neupert (Austin: University of Texas Press, 1992).

Austin, Thomas, "Temporality in Kaurismäki: Anachronism, Allusion, Tableau," in *The Films of Aki Kaurismäki: Ludic Engagements*, ed. by Thomas Austin (London: Bloomsbury Academic, 2018), pp. 17–38.

Bakhtin, Mikhail, *Problems of Dostoyevsky's Poetics*, ed. and trans. by Caryl Emerson (Minneapolis: University of Minnesota Press, 1984).

Bakhtin, Mikhail, *The Dialogic Imagination: Four Essays*, trans. by Caryl Emerson and Michael Holquist (Austin: University of Texas Press, 1982).

Barrier, Michael, *Hollywood Cartoons: American Animation in Its Golden Age* (New York: Oxford University Press, 1999).

Barthes, Roland, "Myth Today," in *Mythologies* (London: Paladin, 1973), pp. 109–64.

Barthes, Roland, "Ornamental Cookery," in *Mythologies* (New York: The Noonday Press, 1972), pp. 78–80.

Baudrillard, Jean, "Simulacra and Simulations," in *Selected Writings*, ed. by Mark Poster (Stanford: Stanford University Press, 1988), pp. 166–84.

Baxter, John, *Buñuel* (New York: Carroll & Graf, 1994).

Benjamin, Walter, "The Author as Producer," in *Understanding Brecht*, trans. by Anna Bostock (London: Verso, 1998), pp. 85–104.

Benjamin, Walter, "The Storyteller," in *Illuminations*, ed. by Hannah Arendt, trans. by Harry Zohn (London: Jonathan Cape, 1970), pp. 83–110.

Berger, John, *Ways of Seeing* (Harmondsworth: Penguin, 1972).

Berlin, Isaiah, *Four Essays on Liberty* (Oxford: Oxford University Press, 1969).

Bettelheim, Bruno, *Freud and Man's Soul* (London: Fontana, 1985).

Beumers, Birgit, *A History of Russian Cinema* (New York: Berg, 2009).

Bloom, Paul, *Against Empathy: The Case for Rational Compassion* (New York: Ecco, 2016).

Boddy, Kasia, "Short Cuts and Long Shots: Raymond Carver's Stories and Robert Altman's Film," *Journal of American Studies*, 34.1 (2000), 1–22.

Bordwell, David, "The Art Film as a Mode of Film Practice," in *Film Theory and Criticism: Introductory Readings*, ed. by Leo Braudy and Marshall Cohen, 5th edn (New York: Oxford University Press, 1999), pp. 716–24.

Bordwell, David, *Making Meaning: Inference and Rhetoric in the Interpretation of Cinema* (Cambridge, MA: Harvard University Press, 1989).

Bordwell, David, *Narration in the Fiction Film* (Madison: University of Wisconsin Press, 1985).

Bordwell, David, "The Classical Hollywood Style, 1917–1960," in *The Classical Hollywood Cinema: Film Style and Mode of Production to 1960*, ed. by David Bordwell, Janet Staiger, and Kristin Thompson (London: Routledge, 1988), pp. 1–86.

Brandon, Ruth, *Surreal Lives: The Surrealists 1917–1945* (London: Macmillan, 1999).

Brooks, Peter, *The Melodramatic Imagination: Balzac, Henry James, Melodrama, and the Mode of Excess* (New York: Columbia University Press, 1985).

Buñuel, Luis, *My Last Sigh,* trans. by Abigail Israel (New York: Vintage, 1983).

Cavell, Stanley, "The World Viewed," in *Film Theory and Criticism: Introductory Readings*, ed. by Leo Braudy and Marshall Cohen (New York: Oxford University Press, 1999), pp. 334–44.

Cawelti, John G, "Chinatown and Generic Transformation in Recent American Films," in *Film Theory and Criticism: Introductory Readings*, ed. by Gerald Mast and Marshall Cohen, 3rd edn (New York: Oxford University Press, 1985), pp. 503–19.

Chanady, Amaryll, *Magical Realism and the Fantastic: Resolved Versus Unresolved Antinomy* (New York: Garland, 1985).

Cocks, Geoffrey, "Death by Typewriter: Stanley Kubrick, the Holocaust, and the Shining," in *Stanley Kubrick: Film and the Uses of History*, ed. by Geoffrey Cocks, James Diedrick, and Glenn Perusek (Madison: The University of Wisconsin Press, 2006), pp. 185–217.

Cocks, Geoffrey, Diedrick, James, and Perusek, Glenn, "Introduction: Deep Focus," in *Depth of Field: Stanley Kubrick, Film and the Uses of History*, ed. by Geoffrey Cocks, James Diedrick, and Glenn Perusek (Madison: The University of Wisconsin Press, 2006), pp. 3–30.

Code, David J, "Don Juan in Nadsat: Kubrick's Music for *A Clockwork Orange*," *Journal of the Royal Musical Association*, 139.2 (2014), 339–86.

Cohen, Alain J., "Stochastics of Sex and Death in Basic |Filmic| Instinct," *Semiotica*, 112.1–2 (1996), 109–21.

Condee, Nancy, *The Imperial Trace: Recent Russian Cinema* (Oxford: Oxford University Press, 2009).

Conley, Tom, "Buñuel Entomographer: From Las Hurdes to Robinson Crusoe," in *A Companion to Luis Buñuel*, ed. by Rob Stone and Julián Daniel Gutiérrez-Albilla (Chichester: Wiley-Blackwell, 2013), pp. 188–202.

Cook, David A., "American Horror: The Shining," *Literature/Film Quarterly*, 12.1 (1984), 2–4.

Cooper, Sarah, "Surreal Souls: *Un Chien Andalou* and Early French Film Theory," in *A Companion to Luis Buñuel*, ed. by Rob Stone and Julián Daniel Gutiérrez-Albilla (Chichester: Wiley-Blackwell, 2013), pp. 141–55.

Cousins, Mark, *The Story of Film* (London: Pavilion Books, 2004).

Danto, Arthur C, *The Philosophical Disenfranchisement of Art* (New York: Columbia University Press, 1986).

DiPaola, Peter Daniel, "The Criminal Time Bomb: An Examination of the Effect of the Russian Mafia on the Newly Independent States of the Former Soviet Union," *Indiana Journal of Global Legal Studies*, 4.1 (Fall 1996), 145–81.

Dobrenko, Evgeny, *Stalinist Cinema and the Production of History: Museum of the Revolution* (Edinburgh: Edinburgh University Press, 2008).

Dunlop, John B, "Russian Nationalist Themes in Soviet Film of the 1970s," in *The Red Screen: Politics, Society, Art in Soviet Cinema*, ed. by Anna Lawton (London: Routledge, 1992).

Dunne, Michael, "*Wild at Heart* Three Ways: Lynch, Gifford, Bakhtin," *Literature/Film Quarterly*, 23.1 (1995), 6–13.

Durbach, Nadja, *The Spectacle of Deformity: Freak Shows and Modern British Culture* (Berkeley: University of California Press, 2009).

Edwards, Gwynne, *A Companion to Luis Bunuel* (Woodbridge: Tamesis, 2005).

Elsaesser, Thomas, *Fassbinder's Germany: History, Identity, Subject* (Amsterdam: Amsterdam University Press, 1996).

Evans, Peter William, *The Films of Luis Buñuel: Subjectivity and Desire* (Oxford: Oxford University Press, 1995).

Fan, Victor, "Redressing the Inaccessible through the Re-Inscribed Body: *In a Year with 13 Moons* and Almodóvar's *Bad Education*," in *A Companion to Rainer Werner Fassbinder*, ed. by Brigitte Peucker (Chichester: Wiley-Blackwell, 2012), pp. 118–41.

Fehrenbach, Heide, *Cinema in Democratizing Germany* (Chapel Hill: The University of North Carolina Press, 1995).

Fijalkowski, Krzysztof, and Richardson, Michael (eds), "The Platform of Prague Twenty Years On," in *Surrealism against the Current: Tracts and Declarations* (London: Pluto Press, 2001), pp. 81–92.

Gabbard, Glen O, and Gabbard, Krin, *Psychiatry and the Cinema* (Washington, DC: American Psychiatric Association, 1999).

Gabbard, Krin, and Sharma, Shailaja, "Stanley Kubrick and the Art Cinema," in *Stanley Kubrick's: A Clockwork Orange*, ed. by Stuart Y. McDougal (Cambridge: Cambridge University Press, 2003), pp. 85–108.

Gerould, Daniel, "Russian Formalist Theories of Melodrama," in *Imitations of Life: A Reader on Film, Television and Melodrama*, ed. by Marcia Landy (Detroit: Wayne University Press, 1991), pp. 121–29.

Glass, Fred, "The 'New Bad Future': RoboCop and 1980s' Sci-Fi Films," *Science as Culture*, 1.5 (1989), 7–49.

Gledhill, Christine, "Introduction," in *Home Is Where the Heart Is: Studies in Melodrama and the Woman's Film*, ed. by Christine Gledhill (London: British Film Institute, 1987), pp. 1–4.

Godwin, George K., "Review: *Eraserhead* by David Lynch, *Film Quarterly*," 39.1 (1985), 37–43.

Golovskoy, Val, "Art and Propaganda in the Soviet Union, 1980–5," in *The Red Screen: Politics, Society, Art in Soviet Cinema*, ed. by Anna Lawton (London: Routledge, 1992), pp. 263–74.

Gorfinkel, Elena, "Impossible, Impolitic: Ali: Fear Eats the Soul and Fassbinder's Asynchronous Bodies," in *A Companion to Rainer Werner Fassbinder*, ed. by Brigitte Peucker (Chichester: Wiley-Blackwell, 2012), pp. 502–14.

Griffiths, Kate, "Spectral Cinema: Le Journal d'une femme de chamber," in *A Companion to Luis Buñuel*, ed. by Rob Stone and Julián Daniel Gutiérrez-Albilla (Chichester: Wiley-Blackwell, 2013), pp. 394–413.

Guback, Thomas, *The International Film Industry: Western Europe and America since 1915* (Bloomington: Indiana University Press, 1969).

Hanstein, Ulrike, "Masquerading, Underacting and Screen Performances," in *The Films of Aki Kaurismäki: Ludic Engagements*, ed. by Thomas Austin (London: Bloomsbury Academic, 2018), pp. 169–86.

Hashamova, Yana, "Aleksei Balabanov's Russian Hero: Fantasies of Wounded National Pride," *The Slavic and East European Journal*, 51.2, Special Forum Issue: "Resent, Reassess, and Reinvent: The Three R's of Post-Soviet Cinema" (Summer, 2007), 295–311.

Hay, Jimmy, "Inside/Outside: Space and Sexual Behavior in Belle de jour and La Pianiste," in *A Companion to Luis Buñuel*, ed. by Rob Stone and Julián Daniel Gutiérrez-Albilla (Chichester: Wiley-Blackwell, 2013), pp. 554–71.

Hayman, Ronald, *Fassbinder: Film Maker* (London: Weidenfeld and Nicolson, 1984).

Highet, Gilbert, *The Anatomy of Satire* (Princeton: Princeton University Press, 1962).

Hoffman, Karen D, "Where the Rainbow Ends: *Eyes Wide Shut*," in *The Philosophy of Stanley Kubrick*, ed. by Jerold J Abrams (Lexington: The University Press of Kentucky, 2007), pp. 59–84.

Hopkins, David, *Dada and Surrealism: A Very Short Introduction* (Oxford: Oxford University Press, 2004).

Hutcheon, Linda, *A Theory of Parody: The Teachings of Twentieth Century Art Forms* (New York: Methuen, 1985).

Hutcheon, Linda, *Irony's Edge: The Theory and Politics of Irony* (London: Routledge, 1994).

Jameson, Frederic, "Postmodernism and Consumer Culture," in *Postmodernism and Its Discontents: Theories, Practices*, ed. by Anne E. Kaplan (London: Verso, 1988), pp. 13–29.

Jones, Julie, "Luis Buñuel and the Politics of Self-Presentation," in *A Companion to Luis Buñuel*, ed. by Rob Stone and Julián Daniel Gutiérrez-Albilla (Chichester: Wiley-Blackwell, 2013), pp. 79–97.

Kaapa, Pietari, "Three Ecologies of Kaurismäki," in *The Films of Aki Kaurismaki: Ludic Engagements*, ed. by Thomas Austin (London: Bloomsbury Academic, 2018), pp. 39–56.

Katz, Robert, *Love Is Colder than Death: The Life and Loves of Rainer Werner Fassbinder* (New York: Random House, 1987).

Kelly, Andrew, "The Brutality of Military Incompetence: *Paths of Glory* (1957)," *Historical Journal of Film, Radio and Television*, 13.2 (1993), 215–27.

Kern, Stephen, *Modernism after the Death of God* (London: Routledge, 2017).

Kerr, Walter, *The Silent Clowns* (New York: Knopf, 1975).

Kinder, Marsha, "The Road and the Room: Narrative Drive in the Films of Luis Buñuel," in *A Companion to Luis Buñuel*, ed. by Rob Stone and Julián Daniel Gutiérrez-Albilla (Chichester: Wiley-Blackwell, 2013), pp. 431–54.

King, Jamie, "Bug Planet: The Frontier Myth of *Starship Troopers*," *Futures*, 30.10 (1998), 1017–26.

Kreimeier, Klaus, *The UFA Story*, trans. by Robert Kimber and Rita Kimber (New York: Hill and Wang, 1996).

Leonard, Philip, "Teach Phenomenology the Bomb: *Starship Troopers*, the Technologized Body, and Humanitarian Warfare," *European Journal of American Culture*, 25.1 (2005), 31–46.

Loukides, Paul, "Taking Inventory: Stock Elements of American Feature Films," in *Symbiosis: Popular Culture and Other Fields*, ed. by Ray R. Browne and Marshal W. Fishwick (Bowling Green: Bowling Green State University Popular Press, 1988), pp. 52–56.

Lynch, David, and Rodley, Chris, *Lynch on Lynch*, revised edn (New York: Faber and Faber, 2005).

MacDowell, James, *Irony in Film* (London: Palgrave Macmillan, 2016).

Marshall, Herbert, "The New Wave in Soviet Cinema," in *The Red Screen: Politics, Society, Art in Soviet Cinema*, ed. by Anna Lawton (London: Routledge, 1992), pp. 173–90.

Matz, Aaron, *Satire in the Age of Realism* (Cambridge: Cambridge University Press, 2010).

McGowan, Todd, *The Impossible David Lynch* (New York: Columbia University Press, 2007).

McMahon, Patrick, "Robert Altman's Cinema of Subversion," *Film Matters*, 2.1 (Spring 2011), 21–25.

Miller, Mark Crispin, "2001: A Cold Descent," in *Depth of Field: Stanley Kubrick, Film and the Uses of History*, ed. by Geoffrey Cocks, James Diedrick, and Glenn Perusek (Madison: The University of Wisconsin Press, 2006), pp. 122–45.

Monaco, James, *How to Read a Film: The World of Movies, Media, and Multimedia* (New York: Oxford University Press, 2000).

Morson, Gary Saul, "Parody, History, and Metaparody," in *Rethinking Bakhtin: Extensions and Challenges*, ed. by Gary Saul Morson and Caryl Emerson (Evanston: Northwestern University Press, 1989), pp. 63–86.

Mortimore, Roger, "Buñuel, Sáenz de Heredia, and Filmófono," *Sight and Sound*, 44 (Summer 1975), 180–82.

Nestingen, Andrew, *The Cinema of Aki Kaurismäki: Contrarian Stories* (New York: Wallflower Press, 2013).

Nilsson, Johan, *American Film Satire in the 1990s* (New York: Palgrave Macmillan, 2013).

Norega, Chon, "A Whisper of Satire (Round Table: Showgirls)," *Film Quarterly*, 56.3 (Spring 2003), 36–38.

Olson, Greg, *David Lynch: Beautiful Dark* (Lanham: The Scarecrow Press, 2008).

Palmer, R Barton, "*The Shining* and Anti-Nostalgia: Postmodern Notions of History," in *The Philosophy of Stanley Kubrick*, ed. by Jerold J Abrams (Lexington: The University Press of Kentucky, 2007), pp. 201–20.

Parciack, Ronie, "The World as Illusion: Rediscovering *Mulholland Dr.* and *Lost Highway* through Indian Philosophy," in *The Philosophy of David Lynch*, ed. by William J Devlin and Shai Biderman (Lexington: The University Press of Kentucky, 2011), pp. 77–93.

Perusek, Glenn, "Kubrick's Armies: Strategy, Hierarchy, and Motive in the War Films of Stanley Kubrick," in *Depth of Field: Stanley Kubrick, Film and the Uses of History*, ed. by Geoffrey Cocks, James Diedrick, and Glenn Perusek (Madison: The University of Wisconsin Press, 2006), pp. 77–100.

Peucker, Brigitte, "In *Despair*: Performance, Citation, Identity," in *A Companion to Rainer Werner Fassbinder*, ed. by Brigitte Peucker (Chichester: Wiley-Blackwell, 2012), pp. 290–312.

Raghavendra, MK (ed.), "A Trajectory of Form: The Development of Soviet/Russian Cinema (1910–2010)," in *Locating World Cinema: Interpretations of Film as Culture* (New Delhi: Bloomsbury, 2020), pp. 165–210.

Rhode, Eric, "Why Neorealism Failed," *Sight and Sound*, 30.1 (Winter 1960), 27–32.

Rice, Julian, *Kubrick's Hope: Discovering Optimism from 2001 to Eyes Wide Shut* (Lanham: The Scarecrow Press, 2008).

Richards, Arlene Kramer, "Woman as Medusa in Basic Instinct," *Psychoanalytic Inquiry: A Topical Journal for Mental Health Professionals*, 18.2 (2009), 269–80.

Richardson, Michael, *Surrealism and Cinema* (New York: Berg, 2006).

Riches, Simon, "Intuition and Investigation into Another Place: The Epistemological Role of Dreaming in *Twin Peaks* and Beyond," in *The Philosophy of David Lynch*, ed. by William J Devlin and Shai Biderman (Lexington: The University Press of Kentucky, 2011), pp. 25–44.

Roberts, Benjamin B, *Sex, Drugs and Rock and Roll in the Dutch Golden Age* (Amsterdam: Amsterdam University Press, 2017).

Robertson, Michael, "Property and Privatisation in *RoboCop*," *International Journal of Law in Context*, 4.3 (2008), 217–35.

Rodríguez, María Pilar, "*Susana*: Melodrama and the Voluptuosity of Destruction," in *A Companion to Luis Buñuel*, ed. by Rob Stone and Julián Daniel Gutiérrez-Albilla (Chichester: Wiley-Blackwell, 2013), pp. 240–54.

Rollin, Roger B, "Robert Altman's *Nashville*: Popular Film and the American Character," *South Atlantic Bulletin*, 42.4 (1977), 41–50.

Rosenbaum, Jonathan, "In Dreams Begin Responsibilities," in *Depth of Field: Stanley Kubrick, Film and the Uses of History*, ed. by Geoffrey Cocks, James Diedrick, and Glenn Perusek (Madison: The University of Wisconsin Press, 2006), pp. 245–54.

Rotellar, Manuel, "Luis Buñuel en Filmófono," *Cinema 2002*, 37 (1978), 36–40.

Scholes, Robert, "Narration and Narrativity in Film," in *Film Theory and Criticism: Introductory Readings*, ed. by Gerald Mast and Marshall Cohen, 3rd edn (New York: Oxford University Press, 1985), pp. 390–404.

Seppala, Jaakko, "The Camera's Ironic Point of View: Notes on the Strange and Comic Elements in the Films of Aki Kaurismäki," in *The Films of Aki Kaurismaki: Ludic Engagements*, ed. by Thomas Austin (London: Bloomsbury Academic, 2018), pp. 83–100.

Shlapentokh, Dmitry, "Drunkenness and Anarchy in Russia: A Case of Political Culture," *Russian History*, 18.4 (Winter 1991), 457–500.

Sontag, Susan, "Against Interpretation," in *A Susan Sontag Reader* (Harmondsworth: Penguin, 1982), pp. 95–104.

Stam, Robert, *Reflexivity in Film and Literature: From Don Quixote to Jean-Luc Godard* (New York: Columbia University Press, 1992).

Stark, Craig, "'What, Me Worry?': Teaching Media Literacy through Satire and Mad Magazine," *The Clearing House: A Journal of Educational Strategies, Issues and Ideas*, 76.6 (2003), pp. 305–09.

Stauffacher, Frank, *Art in Cinema* (San Francisco: San Francisco Museum of Art, 1947).

Tan, Ed S, *Emotion and the Structure of Narrative Film: Film as an Emotion Machine,* trans. by Barbara Fasting (New York: Routledge, 2011).

Thomas, Paul, "Fassbinder: The Poetry of the Inarticulate," *Film Quarterly*, 30.2 (Winter 1976–77), pp. 2–17.

Thompson, David (ed.), *Altman on Altman* (London: Faber and Faber, 2006).

Timofeevsky, Alexander, "The Last Romantics," in *Russian Critics on the Cinema of Glasnost*, ed. by Michael Brashinsky and Andrew Horton (Cambridge: Cambridge University Press, 1994), pp. 24–29.

Verstraten, Peter, "The Irony of Irony: On Paul Verhoeven's Adaptation of Gerard Reve's *De vierde man* [The Fourth Man]," *Journal of Dutch Literature*, 4.2 (2013), 27–42.

Vidal, Agustín Sánchez, "*L'Âge d'or*," in *A Companion to Luis Buñuel*, ed. by Rob Stone and Julián Daniel Gutiérrez-Albilla (Chichester: Wiley-Blackwell, 2013), pp. 172–87.

Warnes, Christopher, *Magical Realism and the Post-Colonial Novel: Between Faith and Irreverence* (New York: Palgrave Macmillan, 2009).

Warshow, Robert, "Movie Chronicle: The Westerner," in *Film Theory and Criticism: Introductory Readings*, ed. by Gerald Mast and Marshall Cohen, 3rd edn (New York: Oxford University Press, 1985), pp. 434–50.

West, Darrell M., and Orman, John M., *Celebrity Politics: Real Politics in America* (Upper Saddle River: Prentice Hall, 2003).

Williams, FD, "The Morality of Bonnie and Clyde," *Journal of Popular Culture*, 4.1 (Summer 1970), 299–307.

Williams, Linda, *Figures of Desire: A Theory and Analysis of Surrealist Film* (Berkeley: University of California Press, 1981).

Wilson, John K, *The Myth of Political Correctness* (Durham: Duke University Press, 1995).

Zizek, Slavoj, *The Art of the Ridiculous Sublime: On David Lynch's Lost Highway* (Seattle: University of Washington Press, 2000).

Zuckoff, Mitchel, *Robert Altman: The Oral Biography* (New York: Vintage, 2010).

Index

www.ingramcontent.com/pod-product-compliance
Ingram Content Group UK Ltd.
Pitfield, Milton Keynes, MK11 3LW, UK
UKHW052112200725
460943UK00003B/3

9 781835 951408